Ecologies of Gender

Ecologies of Gender: Contemporary Nature Relations and the Nonhuman Turn examines the role of gender in recent debates about the nonhuman turn in the humanities, and critically explores the implications for a contemporary theory of gender and nature relations.

The interdisciplinary contributions in this volume each provides theoretical reflections based on an analysis of specific naturecultural processes. They reveal how "ecologies of gender" are constructed through aesthetic, epistemological, political, technological and economic practices that shape multispecies and material interrelations as well as spatial and temporal orderings. The volume includes contributions from cultural anthropology, cultural studies, film studies, literary studies, media studies, philosophy and theatre studies. The essays are organized around four key dimensions of an "ecological" understanding of gender: "creatures", "materials", "spaces" and "temporalities".

The overall aim of the volume *Ecologies of Gender: Contemporary Nature Relations and the Nonhuman Turn* is to explore the potentialities and limitations of the nonhuman turn for a critical analysis and theory of ecologies of gender, and thereby make an original contribution to both the environmental humanities and gender studies.

This book will be of great interest to scholars and students from the interdisciplinary field of the environmental humanities and environmental studies more broadly, as well as from gender studies and cultural theory.

Susanne Lettow is a Senior Researcher at the Margherita von Brentano Center for Gender Studies and teaches at the Institute of Philosophy, Freie Universitat Berlin, Germany.

Sabine Nessel is a Professor for Film Studies at the Institute of Theatre Studies at the Freie Universitat Berlin, Germany.

Routledge Environmental Humanities

Series editors: Scott Slovic (University of Idaho, USA), Joni Adamson (Arizona State University, USA) and Yuki Masami (Aoyama Gakuin University, Japan)

The *Routledge Environmental Humanities* series is an original and inspiring venture recognising that today's world agricultural and water crises, ocean pollution and resource depletion, global warming from greenhouse gases, urban sprawl, overpopulation, food insecurity and environmental justice are all *crises of culture.*

The reality of understanding and finding adaptive solutions to our present and future environmental challenges has shifted the epicenter of environmental studies away from an exclusively scientific and technological framework to one that depends on the human-focused disciplines and ideas of the humanities and allied social sciences.

We thus welcome book proposals from all humanities and social sciences disciplines for an inclusive and interdisciplinary series. We favour manuscripts aimed at an international readership and written in a lively and accessible style. The readership comprises scholars and students from the humanities and social sciences and thoughtful readers concerned about the human dimensions of environmental change.

For more information about this series, please visit: www.routledge.com/Routledge-Environmental-Humanities/book-series/REH

Ecologies of Gender

Contemporary Nature Relations and the Nonhuman Turn

Edited by
Susanne Lettow and Sabine Nessel

Routledge
Taylor & Francis Group
LONDON AND NEW YORK

First published 2022
by Routledge
4 Park Square, Milton Park, Abingdon, Oxon OX14 4RN

and by Routledge
605 Third Avenue, New York, NY 10158

Routledge is an imprint of the Taylor & Francis Group, an informa business

British Library Cataloguing-in-Publication Data
A catalogue record for this book is available from the British Library

Library of Congress Cataloging-in-Publication Data
A catalog record has been requested for this book

ISBN: 978-0-367-90239-1 (hbk)
ISBN: 978-1-032-19831-6 (pbk)
ISBN: 978-1-003-02331-9 (ebk)

DOI: 10.4324/9781003023319

Typeset in Times New Roman
by codeMantra

Contents

Figures

Contributors

Sven Bergmann is a Cultural Anthropologist working at the German Maritime Museum in Bremerhaven. His research interests span from feminist technoscience, kinship, reproduction, and the body to political ecology, marine and environmental anthropology, and economies of waste. His current research focuses on maritime environmental phenomena that are often unpredictable and speculative.

Angelica Fenner is Associate Professor of German and Cinema Studies at the University of Toronto. She is author of *Race under Reconstruction in German Cinema* (University of Toronto Press, 2011), co-editor of *The Autobiographical Turn in Germanophone Documentary and Experimental Film* (Camden House, 2014) and guest-editor of special issues of *Transit* (2014), *Camera Obscura* (2018) and *Feminist German Studies* (2022).

N S Gundur is Professor at the Department of Studies and Research in English, Tumkur University (India). He is a bilingual writer whose areas of interest include literary and cultural theory, critical European thought, liberal arts education and Kannada studies. He has translated Michel Foucault's writings into Kannada and is the author of *Vartamaanada Itihaasakaara: Michel Foucault* (Kuvempu Bhasha Bharathi Pradhikara, 2017), a critical introduction to Foucault in Kannada.

Susanne Lettow is Senior Researcher at the Margherita-von-Brentano Centre for Gender Studies and teaches at the Institute for Philosophy at Freie Universität Berlin. Her publications include *Conceiving Reproduction in German Naturphilosophie*, a topical edition of the journal *History and Philosophy of the Life Sciences* (2021), co-edited with Gregory Rupik, *Race, Gender and Reproduction in Philosophy and the Early Life Sciences* (SUNY Press, 2014) and *Biophilosophien. Wissenschaft, Technologie und Geschlecht im philosophischen Diskurs der Gegenwart.* [*Biophilosophies. Science, Technology and Gender in the Philosophical Discourse of the Present*] (Campus 2011).

Natania Meeker is Associate Professor of French and Comparative Literature at the University of Southern California. Together with Antónia Szabari she co-authored *Radical Botany: Plants and Speculative Fiction* (Fordham University Press, 2019) and she is the winner of the Science Fiction and Technoculture Studies Book Prize. Meeker's interests include feminist theory, the history of materialisms and eighteenth-century literature and philosophy. She is currently working on a monograph tentatively entitled *Illusion Without Error*, which examines the enmeshment of Enlightenment materialist philosophy with femininity, literal and figural.

Ramona Mosse is the Principal Investigator of the Volkswagen Foundation-funded research project *Viral Theatres: Performing Post/Pandemic Cultures in the Anthropocene* at the EXC 2020 Temporal Communities, Freie Universität Berlin. Her work has been published in journals such as *Theatre Journal*, *Anglia*, *Thewis* and *Global Performance Studies*. She is co-editor of Erika Fischer-Lichte's *Routledge Introduction to Theatre and Performance Studies* (2014). Ramona Mosse holds a PhD in English and Comparative Literature from Columbia University.

Sabine Nessel is Professor for Film Studies at Freie Universität Berlin. She is the project leader of the research project *Figurations of Difference in Moving Images* (funded by the German Research Foundation DFG) and is a principal investigator of the DFG Research Training Group *Normativity, Critique, Change*. Her research focuses on the areas of film and cinema theory, media cultural animal studies, environmental humanities and film, and gender studies. Sabine Nessel is a founding member and member of the advisory board of the peer-reviewed open access journal *Nach dem Film*.

Sangita Patil is an Assistant Professor. Her areas of interests include ecofeminism, Kannada theatre and cultural studies. She is a short story writer and the author of *Ecofeminism and Indian Novel* (Routledge 2018) and *Reconstructing ecofeminism: A Study of Kamala Markandaya's Nectar in a Sieve* in *Cogent Social Sciences*.

Kathrin Peters is Professor of Visual Culture Studies at the Berlin University of the Arts and co-speaker of the DFG Research Training Group *The Knowledge of the Arts*, also at the Berlin University of the Arts. She is founding member and member of the advisory board of the *Zeitschrift für Medienwissenschaft* (*Journal for Media Studies*). Peters publishes on gender and media, contemporary film and design history.

Swarnalatha Rangarajan is Professor of English at the Department of Humanities and Social Sciences at the Indian Institute of Technology Madras, and founding editor of the *The Indian Journal of Ecocriticism*. Recent works include *Mayilamma: The Life of a Tribal Eco-Warrior* (Orient Blackswan, 2018) which she co-translated with Sreejith Varma and *Lifescapes: Interviews with Contemporary Women Writers from Tamil Nadu* (Women Unlimited, 2019), a book she co-authored with K. Srilata. Together with Scott Slovic and Vidya Sarveswaran she co-edited the forthcoming *The Routledge Handbook of Ecocriticism and Environmental Communication*. Swarnalatha Rangarajan is also an author of short fiction.

Catriona (Cate) Sandilands is Professor of Environmental Arts and Justice at the Faculty of Environmental and Urban Change at York University. Her recent and upcoming works on plants and vegetal relationships include essays in *The Cambridge Companion to the Environmental Humanities,* as well as *Kin: Thinking with Deborah Bird Rose*; *Plant Fever: Towards a Phyto-Centric Design*; *Veer Ecology: Key Words for Ecotheory*; *Sex Ecologies*; *Resilience*; *Women' Studies*; and *GLQ*. She is currently completing the monograph *Plantasmagoria: Botanical Encounters in the (M) Anthropocene*.

Andrea Seier is Professor for Cultural History of Audiovisual Media in the Department for Theatre, Film and Media Studies at Vienna University. Her PhD on gender and media performativities was completed at Ruhr University Bochum, Germany. Research interests include micropolitics of audiovisual media, governmentality and gender studies, as well as class, precarity and social im/mobility in audiovisual media.

Nicole Seymour works at the intersection of environmental studies and queer studies. She is the author of *Bad Environmentalism: Irony and Irreverence in the Ecological Age* (University of Minnesota Press, 2018). In 2022, Bloomsbury's Object Lessons series will publish her cultural history of glitter. She is Associate Professor of English at California State University, Fullerton.

Antónia Szabari is Associate Professor of French and Comparative Literature at the University of Southern California. Her interests include speech acts and queer agencies in early modern literature, the history of botany, vegetal ontologies and speculative fiction. She is the author of *Less Rightly Said: Scandals and Readers in Sixteenth-Century France* (Stanford University Press, 2009) and, with Natania Meeker, *Radical Botany: Plants and Speculative Fiction* (Fordham University Press, 2019).

Introduction

Ecologies of Gender and the Nonhuman Turn

Susanne Lettow and Sabine Nessel

Contemporary nature relations are contested and crisis-afflicted in a paradoxical manner: marked by the effects of climate change, pandemics and extinctions, yet at the same time characterized by a search for new materials and processes that will allow economic growth even after we stop using fossil fuels. While critical discourses and social movements point forcefully to the fragility of terrestrial conditions of existence, the conquest of new micro- and macro-worlds continues apace, from bioeconomic commodification of living beings to exploring the possibilities of space tourism. The causes and effects of ecological devastation and sociopolitical modes of shaping social nature relations are structured by a host of intersecting power relations and differences. Ecofeminist theories and analyses in the field of gender and environment studies have shed light on many facets of the intersectional constellations, in which gender and class relations, geopolitical and regional hierarchies rooted in a long colonial history and practices and cultural constructs of human exceptionalism are interwoven. They have also, in various ways, drawn attention to how horizons of thought and imagination produced by theoretical and aesthetic practices play a constitutive role in the formation of gender and nature relations. Historically evolved relations of power and domination, and the conflicts associated with them, are congealed within social images and narratives of human and nonhuman nature, gender, sexuality and generativity. At the same time, images and narratives are crucial points of disembarkation for developing new modes of thought and existence.

Proceeding from this insight, the present volume aims to investigate theoretical and aesthetic depictions and narratives of gender and nature relations in literature, film, art, theatre, philosophy and science and technology studies, and thereby contribute to a critical theory of ecologies of gender. As the essays collected in this volume show, concepts and ideas of gender, nature, human, animal, plant and life that are historically embedded in long traditions of anthropocentric-extractivist and patriarchal ways of seeing and thinking need to be reconceptualized and reimagined in the light of crisis-afflicted nature relations.

DOI: 10.4324/9781003023319-1

Crucial input and inspiration for this project can be drawn from the *nonhuman turn* in cultural studies and the humanities (see, e.g. Braidotti 2013, Grusin 2017), which fundamentally questions human exceptionalism and has prompted an engagement with nonhuman phenomena. Anthropocentric understandings of history and Western theories of metaphysics, as well as the systems that produce them, have been subjected to intense scrutiny and reformulation, especially in the transdisciplinary fields of human and cultural animal studies, critical plant studies and new materialism. But the theoretical implications of the nonhuman turn have also increasingly been a topic of critical discussion. For instance, Catriona Sandilands observes that "some recent work under the broad (and polyvocal) heading of 'material' feminism has tended to downplay [...] ethical and political concerns as part of a turn to an ontologically driven inquiry" (2017, 446). Dana Luciano and Mel Y. Chen similarly speak of the "politically limiting effects" (2015, 194) of the nonhuman turn, and several authors have suggested that the departure from the epistemological paradigm of "situated knowledge" (Deuber-Mankowsky 2019) is problematic. This volume builds on these critical analyses, which – like the nonhuman turn in general – provide key theoretical underpinnings for the emerging field of the environmental humanities.

Central to this interdisciplinary field, which has flourished over the past decade, is "an effort to enrich environmental research with a more extensive conceptual vocabulary, whilst at the same time vitalising the humanities by rethinking the ontological exceptionality of the human" (Rose et al. 2012, 2). In this context, ecofeminist, queer and intersectional approaches have played a crucial role from the very start, or to put it another way: interdisciplinary scholarship, which since the late 1980s has been carried out under headings such as ecofeminism, queer ecology or gender and environment studies, is a central foundation on which the environmental humanities build. However, feminist authors have also raised questions about the way in which critical perspectives on gender relations have hitherto been incorporated into the field. Jennifer Hamilton and Astrid Neimanis, for example, have observed a problematic trend "whereby key feminist scholars or concepts may be mentioned, but their feminist investments are not incorporated as such" (2018, 501). As a consequence, these concepts and theoretical insights are often "neutralized" in ways that "undermin[e] the potential for environmental humanities to build alternative worlds" (ibid.). Certainly, feminist and queer approaches have blind spots too, and, as Sharlene Mollett has argued, "the mutual entanglements of gender and race" (Mollett 2017, 155) remain understudied in environmental studies. The essays in this volume therefore adopt intersectional perspectives that highlight the interrelations between multiple forms of difference, power and domination in order to develop a critical and contextualized understanding of gender and nature relations. Accordingly, the authors do not advocate a unitary, coherent understanding of gender and nature relations but write from different angles and perspectives. The essays should thus be understood as situated analyses that are, each in their own way, shaped by the power relations in which they are embedded and with which they engage. Nevertheless, despite dealing with textual and aesthetic materials from various historical and geographical contexts, they converge in opening up a new horizon in which notions of gender, the human and the nonhuman world can be reconceptualized.

Although gender is certainly not the only notion that has been scrutinized in the context of the environmental humanities and the nonhuman turn, we think that the theoretical challenges this concept poses are particularly complex and multilayered. First, the notion is imbued with the problem of false universalism and, in this respect, it is epistemologically similar to the concept of the human. Second, both the sex/gender distinction and its deconstruction through the concept of gender have epistemologically downplayed the "natural" or "organic" aspects of human bodies as well as their interrelations with nonhuman creatures and materials. However, as Stacy Alaimo puts it, these theories "are haunted by the pernicious notions of nature that propel them. Thrust aside, completely removed from culture, this nature – the repository of essentialism and stasis – nonetheless remains dangerously intact" (2010, 5). A reconceptualization of gender in light of the nonhuman turn thus has to actively engage with and rearticulate traditional understandings of nature and the nonhuman world. In order to avoid a vitalist position, in which an all-encompassing concept of "life" masks the historicity and complexity of social relations of power and inequality and the specific conditions and possibilities of human agency, we use the phrase "ecologies of gender". We interpret this expression to mean attending to the systematic interactions and interrelations between human bodies, artefacts and other species, that is, the myriad nonhuman environments and milieus shaped by gender relations. Accordingly, the concept of ecology expands our understanding of situatedness by focussing attention on the dynamic structures of social nature relations. The idea of ecologies in the plural – that is, of different, intersecting relational levels – is important; we understand ecologies as complex, multilayered meshes of relations that are constituted by diverse dynamics, activities and practices and hence are also alterable. "An ecology of practices", writes Isabelle Stengers, "does not have any ambition to describe practices 'as they are' [...] but as they may become" (2005, 186). Consequently, gender relations can be understood as multi-relational orders in which self-relations, relations to the body and intersubjective and social relations always also constitute and are co-constituted by relations to the non-human world of substances, artefacts and other organisms. On this understanding, gender relations are always at the same time nature relations, and nature relations must conversely be conceived as mediated by gender relations. To move past a one-dimensional concept of the environment that posits a clear boundary between an individual and their surroundings – the idea of "organisms plus environment" (Haraway 2016, 30) – Donna Haraway introduced the notions of sympoiesis and tentacularity, which are likewise based on the recognition that relations to self, world and environment need to be constantly reproduced in new, different forms, and always involve a plurality of humans, non-human organisms, substances and artefacts.

The essays collected in this volume are organized around four key dimensions of an "ecological" understanding of gender as described above: "creatures", "materials", "spaces" and "temporalities". The volume includes contributions from literary studies, media studies, film and theatre studies, science and technology studies and philosophy that each reveal central aspects of these four dimensions.

The first section, *Creatures: The Biopolitics of Making Kin*, explores how the interrelations between human bodies and the nonhuman world of living beings can be reconceptualized. The essays in this section build on insights offered by multispecies perspectives and the plant turn in the humanities, which have revealed that human bodies are always already entangled with and co-constituted through a wide range of other creatures. "To be one is always to *become with* many" (Haraway 2008, 4) remarks Haraway, while emphasizing that it is crucial to take into account the

> situated histories, situated naturecultures, in which all the actors become who they are *in the dance of relating*, not from scratch, not ex nihilo, but full of the patterns of their sometimes joined, sometimes separate heritages both before and lateral to *this* encounter.
>
> (Haraway, 25)

Practices and ideas of connectedness and of the transgression of boundaries between human and nonhuman creatures thus need to be understood in the context of the historically specific relations of power, domination and extractivism in which they emerge. Accordingly, calls to "make kin" with other creatures also need to actively address the biopolitical and patriarchal legacies and contexts to which these calls relate. As Katharine Dow and Janelle Lamoreaux argue, it is crucial to acknowledge "kinship's darker side" and to make clear "that kin relations are not always strategic engagements or willful encounters. [...] Situated kinmaking is not only about strategic alliances but also about the destructive elements of relations" (Dow and Lamoreaux 2020, 483–484). For a critical understanding of ecologies of gender it is thus crucial to do both: to scrutinize existing practices and narratives through which the entanglements between human and nonhuman creatures have been constituted, conceptualized and imagined, and to envision non-extractivist and emancipatory multispecies correlations. It is with respect to this task that the three chapters in the first section converge.

In her chapter *Mulberry Intimacies and the Sweetness of Kinship, Catriona Sandilands* scrutinizes the "biopolitics of mulberry–human relations" in southern Ontario by focussing on two mulberry stories through which they are shaped. The first story is one about species identity and the policing of "purity", and deals with "the 'invasion' drama involving *M. alba* and *M. rubra*". The second story deals with the attempts to systematically regulate the sexual expression of mulberries. Sandilands shows that both stories are shaped through contemporary and historical power relations, in particular colonialism, capitalist accumulation and obsessions with purity, utility and excess. While these mulberry intimacies "are not always sites of pleasure" – to say the least – taking seriously the "mulberry logic", which, among other things, undermines a binary understanding of sex, might open up possibilities "for more mutually enriching tendings". *Natania Meeker* discusses the "importance of the plant as a key figure for gender under biopolitics". Her chapter *Vegetal Subjects of Feminist Speculative Fiction* focusses on the writings of Charlotte Perkins Gilman, in particular her feminist utopia *Herland*. By scrutinizing the white supremacist and eugenic implications of Gilman's speculative fiction, Meeker shows that "the turn to the plant does not necessarily mean a turn away from human modes of harm – on the contrary". Meeker argues that Gilman's texts rearticulate earlier biopolitical phantasms and strategies of breeding and cultivating humans, which Meeker discusses with reference to Zygmunt Bauman's critique of modern "gardening culture", that is, a culture fuelled by attempts to forcefully manage and "improve" human and nonhuman nature. With regard to the present plant turn within the humanities, Meeker argues that the "racist biopolitical legacy" of the figure of the plant must not be forgotten but rather critically reworked. While this particularly holds for conceptualizations and ideas of the figure of the plant in modern European cultural and intellectual history, the chapter *The Arboreal Feminine: An Analysis of Affect and Activism in Two Ecofeminist Re-Enchantment Narratives from India* by *Swarnalatha Rangarajan* makes clear that the figure of the (female) plant is not confined to this specific historical and geographic context. Rangarajan analyses two Indian ecofeminist narratives which explore both the critical potential and the violent aspects of this figure. In her readings of *Mother Forest: The Unfinished Story of C. K. Janu* and Sumana Roy's *How I Became a Tree*, Rangarajan explores the ways in which these narratives articulate references to Indian "forest culture", in which the forest is regarded as a "maternal earth body which provides sustenance", to literary traditions "that equate women with flowers and with *prakriti*, the dynamic force of nature" and to political struggles over land and postcolonial development projects. Again, ideas of becoming-plant and of kinship between humans and nonhuman nature appear as complex and ambivalent. However, by discussing these various aspects and legacies of the "arboreal feminine", Rangarajan shows how these texts "challenge narratives of disenchantment" and contribute to a re-enchantment of nature which is crucial for both creative activity and environmental justice activism.

The second section, *Materials: Agency in/of Transcorporeal Assemblages*, investigates how materials are articulated politically, economically, affectively and aesthetically, and what role they play in processes of gendering. The essays in this section build on the idea – formulated in the context of new materialism, which we understand as part of the broader non-human turn – that human bodies are co-constituted by interrelations and interweavements with substances and artefacts, and that the boundaries of bodies are "porous" (Tuana 2008). Recognizing this porous nature of material objects is of key epistemic importance to an understanding of "ecologies of gender"; for instance, if we wish to understand the circulation of toxic substances or of materials such as oil and plastic that form the fundamental building blocks of an industrialized way of life. These substances, alongside pharmaceuticals and other chemicals that are added to human, animal and plant bodies, modify these bodies in specific ways, often to the detriment of their health and the environment. In this context, Michelle Murphy coined the term "chemical regimes of living", remarking that "not only are we experiencing new forms of chemical embodiment that molecularly tie us to local and transnational economies, but so too processed food, hormonally altered meat, and pesticide dependent crops become the material sustenance of humanity's molecular recomposition" (Murphy 2008, 696). Another key insight underpinning the chapters in this section is that substances and materials are not simply given, but always mediated by practices and processes of meaning production. They therefore focus on "material–semiotic" (Haraway) and "transcorporeal" (Alaimo) assemblages, in which agency is configured in a variety of ways. Specific issues that are addressed include the representations of the invisible and unthinkable workings and entanglements of the carbon economy, the cultural and affective significance of materials such as glitter and synthetic testosterone, and the concept of matter itself.

In her chapter *Plastic Ambivalence*, *Nicole Seymour* shows how plastic art, more than simply raising awareness about plastic pollution, serves to articulate our emotional and material embeddedness in this substance. Further, she argues, aestheticizing plastic is a way of living both against and with it. Seymour thus identifies ambivalence as a central affect of this present era, which some now call "the Plasticene". She traces the appearance of this affect through queer ecological thought as well as the contemporary art world – in which found and recycled-plastic artworks have recently proliferated. The chapter concludes with a close look at the plastic-based works of two feminist artists, Ifeoma U. Anyaeji (Nigeria/USA) and Ruth Peché (Spain). While these artists have received less attention than their male counterparts such as Mark Dion, Seymour argues that their use of feminized aesthetic practices offers a crucial commentary on the unevenly distributed violence of plastic pollution. *Kathrin Peters*, in her chapter *Political Drugs: Materiality in* Testo Junkie, discusses Paul B. Preciado's autofiction *Testo Junkie: Sex, Drugs, and Biopolitics in the Pharmacopornographic Era* (2008/2013), which is a document of a self-experiment with testosterone, a discourse history of hormones and a postporno. Against the background of Judith Butler's gender theory, Peters asks whether Preciado conceptualizes "sex" as a material substance of the sexual body, which can be voluntarily manipulated. She concludes that *Testo Junkie,* by exploring processes of subjectivation as male or trans, reveals how somatic events – whether induced by pharmaceuticals or not – always take place within texts, spatial apparatuses and visual artefacts. Based on Preciado's notion of gender ecology the chapter argues that this ecology needs to be understood as an interdependence of technology, gender politics, somato-physical materiality and cultural imagination – and thus as constitutively unstable. In her chapter *Unthinkable Ecologies in Theatres of the Anthropocene, Ramona Mosse* discusses the radical dramaturgical shift that the non-human turn in the humanities has brought to theatre and the development of recent dramatic writing on the environment. Focussing on Ella Hickson's *Oil* (2016) and Duncan Macmillan's *2071* (2014), two plays that share a common concern with playing with temporal and spatial scales to represent the invisible and unthinkable workings and entanglements of the carbon economy, the chapter explores the concept of the "unthinkable" that has shaped the recent environmental theories of Amitav Ghosh, Dipesh Chakrabarty and Donna Haraway. Mosse discusses the question of how the "unthinkable" lends itself to developing new aesthetic forms and narratives that affect how we relate to gender, political community and cultural production.

The third section, *Spaces: Landscapes and Architectures of Power and Imagination*, explores the spatial dimension of ecologies of gender. Gender and nature relations are essentially spatial relations inasmuch as they are always situated in specific geographical and geopolitical arrangements. Moreover, contemporary nature relations and the ecological devastation they are bound up with make abundantly clear that the conditions and effects of the modes of existence through which gender and nature relations are constituted transcend local circumstances and are thus translocal, with a planetary dimension. Like the other dimensions of ecologies, spaces too should be conceived as constituted through material–semiotic practices. Landscapes such as Lusatia – formerly a key coal mining region in eastern Germany and today an area where ecological and socio-economic conflicts merge – or the Sundarbans – an Indian region of mangrove forests – are not only shaped through political histories of extractivism and intersectional power relations but are currently being politically and culturally reappropriated and reshaped, including through literature and film. Such aesthetic practices rework established modes of problematization and open up new ways of thinking. This also holds for aesthetic articulations of built environments, such as "non-places" (Marc Augé) like airports or architecture and interior design. These built environments need to be understood as elements of nature relations through which subjectivities, affects and modes of behaviour are regulated, and which – as in the case of airports – are constitutive of specific modes of acceleration and control over planetary space, and closely correlate with the temporal aspects of extractivism. The essays in this section explore semantic and aesthetic processes of territorialization enacted through film, literature and media. They focus on arrangements in film and literature of space, including processes of extractivism and regionalization.

In her chapter *Gender, Nature, Nonhuman Animal:* Bird People *(2014) and the Proliferation of Difference in Cinema, Sabine Nessel* shows how Donna Haraway's "figures" and Jacques Derrida's "animots" – two key concepts in feminist discourse and cultural animal studies respectively – can be connected to studies of the proliferation of difference in arts and media. Her case study of the film *Bird People,* in which questions of anthropological difference figure prominently, focusses on the interweaving between anthropological, gender and class difference, and proposes a perspective shift on nature relations that opens up an alternative reading of the film. From this perspective, the airport is revealed as a space where differences such as nature/culture and nature/technology become intertwined. In *Bird People,* this ecological perspective on the airport is initiated by a transition from individual gender, anthropological and class differences to a proliferation of difference, in which we can discern the historicity of the idea of individual difference. *Andrea Seier,* meanwhile, explores how the German film *Wild* (2016), directed by Nicolette Krebitz, interweaves the notion of "becoming-wild" with questions of social impositions and expectations and issues of subjectivation, in particular with regard to gender. In her chapter *Wildlife Among Us. Post-Natural Worlds and Interspecies Encounters in Nicolette Krebitz's* Wild, Seier analyses how *Wild* centres around a young woman, Ania, who is "drawn to nature" in a wayward fashion. One day, walking home from work, Ania runs into a full-grown wolf and, after further encounters, starts a kind of relationship with the animal in her apartment, and increasingly rejects the contracts of civilization. The chapter discusses how the film aesthetically articulates the human–animal relationship. Seier's close reading of the film concentrates on the urban environments in which Ania and the wolf interact. The film analysis is framed by an exploration of new materialism and governmentality studies; thus, two distinct approaches from gender studies are brought into a productive dialogue that aids understanding of both the spatial and aesthetic dimensions of ecologies of gender. In her chapter *Creating Emotion with Space in Nanouk Leopold's* Brownian Movement, *Angelica Fenner* investigates the irrepresentability of trauma in visual culture, arguing that the spatialization of affect through film can nonetheless allow trauma's effects to be apprehended indirectly, for example, by creating tensions between trauma's interiority, conceived as psychically and corporeally embedded, and the exteriority of memory, understood as trauma's secondary narrative reworking through artistic work, interpersonal relations or engagement with immediate surroundings. By studying the Belgian arthouse film *Brownian Movement* (dir. Nanouk Leopold, 2010), Fenner explores how a highly haptic mode of eliciting sensory and affective engagement engenders a language for conveying trauma's aftermath. The analysis draws on Derrida's concept of invagination to explicate the spatialization of the female protagonist's implicit trauma as well as the film's textual infolding of notable women characters from literary and film texts authored by the Brothers Grimm, Lewis Carroll, Luis

Buñuel and Alfred Hitchcock. Karen Barad's materialist rethinking of the phenomenological terms of intra-relation provides further scaffolding for unsettling traditional channels of identification with film characters and, by extension, any speculation about causal links between their present actions and potentially unnarrativized or unnarrativizable pasts. In their chapter *Theorizing Ecofeminism through a Spatial Analysis of Amitav Ghosh's* The Hungry Tide, *Sangita Patil and N. S. Gundur* explore the representation of gender and environment through a close examination of Amitav Ghosh's *The Hungry Tide* (2008). Based on an analysis of this novel, the authors re-think the interrelationship between the human and the non-human in a way that blurs the boundaries between these categories and helps to envision an ecohumanist perspective. The chapter discusses the extent to which the environmental humanities need to engage with questions of space, and hence the geopolitics and sociocultural arrangements of particular places. The chapter contributes to ongoing debates on ecofeminism and the non-human turn by deconstructing binary constructions of gender, and by arguing that there is a need for a humanism that transcends narrow views of ecofeminism that focus solely on women. The authors offer a new universalist perspective which they call "ecohumanism".

The fourth section is titled *Temporalities: Histories, Presents and Futures to Change*. The essays in this section deal with temporal aspects of ecologies of gender. Relations to self and body, social relations and relations to nonhuman nature are constituted by relations that are fundamentally temporal in nature. However, the temporal dynamics that shape these different relational levels differ, so that a theory of gender ecologies must attempt to understand the interplay of different temporalities. As Barbara Adam has shown, the dominant focus on linear "industrial time" fundamentally neglects the "proper times" (*Eigenzeiten*), the specific rhythms and dynamics, of the non-human world and interactions between human practices, the activities of other living beings and material processes. "The complex temporalities of the majority of environmental degradations and hazards", writes Adam, "are located outside the reach of this particular conception. That is, a large proportion of the processes associated with the most difficult environmental problems tend to be inaccessible to the senses, invisible until they materialize as symptoms" (Adams 1998, 11). In contemporary debates, the difficulty of perceiving and conceptualizing long-term changes in nature relations and the terrestrial conditions of human, animal and plant life is addressed through the concept of the Anthropocene, which draws a connection between "human" and "geological" time and through the concept of "slow violence". These two concepts are deployed in an attempt to render past and present ecological change intelligible and so allow nature relations to be configured in less destructive ways in future. While the concept of the Anthropocene places the long-problematic figure of the human back at the centre of a grand narrative, the concept of "slow violence" appears to give us a theoretical grasp of the transformations of naturecultural constellations, as well as the intersectional relations of power and violence that structure them. The essays in this section therefore critically interrogate these concepts and explore the ambivalence of present-day visions of the future that are articulated in theories of plant intelligence and works of science fiction.

In her chapter *The Figure of the Human: Philosophical Narratives on Sex, Race and Organic Kinship in the "White (M)anthropocene, Susanne Lettow* scrutinizes "the human" through the lens of history of philosophy. Through a critical evaluation of the emergence of this figure in the late eighteenth and early nineteenth centuries in the context of philosophical anthropology, and rearticulations of this tradition in the first decades of the twentieth century, she shows that the figure of the human has since its inception been shaped by (proto-)biological ideas of organic kinship linked to constructions of sex, race and human exceptionalism. She argues that it is not the universalist aspirations of humanism which are problematic but the fact that these aspirations have been constantly subverted through these constructions. Based on this analysis she argues that Bruno Latour's and Rosi Braidotti's attempts to relegate "the human" back to relations to the earth or to life as such do not seem a critical enough departure from the troubling history of philosophical anthropology. A reconceptualization of the narrative of the Anthropocene should, in contrast, rearticulate the universalist perspective in the sense of Haraway's notion of "non-generic" or "critical" humanity, and at the same time needs to engage in the unfinished process of deconstructing the figure of the human. *Sven Bergmann* explores the concept of slow violence with respect to two cases of environmental problems. His chapter *Speculative Ecologies: Salmon Farming and Marine Microplastics as Slow Disasters* focusses first on toxic algal blooms in southern Chile and second on the accumulation of plastics in the oceans, in this case the ocean around the Polynesian island of Rapa Nui. Based on his own socio-anthropological research on these two cases, Bergmann shows that the slowness of ecological transformations produces a systematic epistemic uncertainty that makes it hard to understand and attribute causes to phenomena of ecological devastation. For political movements that seek to politicize slow violence and the colonial histories it involves, this presents a serious problem. By adopting Heather Davis' notion of queer temporalities, Bergmann argues that uncertainty and ambiguity can also, however, be regarded as epistemically and politically productive. While attempts to reframe phenomena of slow violence in terms of straight time or the "traditional environmental rhetoric" of the doomsday clock, which evoke ideas of purity being destroyed, Bergmann argues for adopting feminist, queer and postcolonial perspectives from science and technology studies that overcome "strict dichotomies between nature and culture, between purity and toxicity". Caring for newly emerging "biological-synthetic entanglements", he concludes, is a more promising strategy for a critical politics of gender ecologies than bemoaning the loss of pristine nature. *Antónia Szabari* explores the ambiguity of future scenarios as they are currently presented in scientific studies on plant intelligence and in speculative fiction. In her chapter *Futures of Plant–Human Mutualism: Science, Technology and Speculative Fiction*, she discusses how futures of a "plant-based bioeconomy", in which the capacities of plants come to serve scientific and biotechnological ends, are imagined and what the role of humans in such a bioeconomy would be. She turns to Sue Burke's Semiosis Duology, comprising the sci-fi novels *Semiosis* and *Interference*, and brings them into conversation with critical plant studies. Szabari shows how these novels construct a fictional world shaped by plant technologies that leave open the question of "whether the result is another story of colonizing [...] or a posthuman mutualism". This ambiguity, Szabari argues, can best be understood in terms of Haraway's phrase "staying with the trouble". While Burke certainly does not portray a positive utopia that we should strive for, she does "hel[p] us come face to face" with distant or not so distant possibilities for new nature relations.

Academic research is never a solo effort, but always the product of a collective endeavour. This book came out of a conference held in November 2019 at Freie Universität Berlin; we would like to thank the German Research Foundation for the generous funding they provided for the conference and to Isa Eichmann and Sophie Holzberger for organizational support. We are also grateful to our colleagues at FU Berlin's Margherita von Brentano Center and Institute of Theater Studies for the institutional support and collaborative working environment they provided while this publication was being prepared. In particular, we would like to extend our thanks to the student assistants Franziska Lisa-Marie Wohlfahrt and Sophie Hartleib, who helped with organizational matters during the production process, and Dorith Budich, who provided invaluable administrative support. We are also indebted to Übersetzungsbüro Denzig and Andrew Godfrey for their highly reliable translation and editing services for some chapters and to interspecies collective CMUK (Clara, Mathias, Ute, Karl), by name Ute Hörner and Mathias Antlfinger, and Ruth Peché for providing photos and reprint permissions. Last but not least, we would like to thank Rosie Anderson and Matthew Shobbrook at Routledge for their dedicated support throughout the production process.

References

Adam, Barbara. 1998. *Timescapes of Modernity. The Environment and Invisible Hazards*. London, New York: Routledge.

Alaimo, Stacy. 2010. *Bodily Natures: Science, Environment, and the Material Self.* Bloomington: Indiana University Press.

Braidotti, Rosi, 2013. *The Posthuman*. Cambridge: Polity Press.

Deuber-Mankowsky, Astrid. 2019. "Das Szenarium, in dem sich Medienanthropologie und Neue Materialismen treffen." In *Medienanthropologische Szenen: Die Conditio Humana im Zeitalter der Medien*, edited by Christiane Voss, Katerina Krtilova and Lorenz Engell, 31–42. Munich: Fink.

Dow, Katharine, and Janelle Lamoreaux. 2020. "Situated Kinmaking and the Population 'Problem'." *Environmental Humanities* 12 (2): 475–491.
Grusin, Richard (ed.). 2017. *Anthropocene Feminism*. Minneapolis: University of Minnesota Press.
Hamilton, Jennifer Mae, and Astrida Neimanis. 2018. "Composting Feminisms and Environmental Humanities." *Environmental Humanities* 10 (2): 501–527.
Haraway, Donna. 2008. *When Species Meet*. Durham. NC: Duke University Press.
Haraway, Donna. 2016. *Staying with the Trouble: Making Kin in the Chthulucene*. Durham, NC: Duke University Press.
Luciano, Dana, and Mel Y. Chen. 2015. "Introduction: Has the Queer Ever Been Human?." *GLQ: A Journal of Lesbian and Gay Studies* 12 (2–3): 183–197.
Mollett, Sharlene. 2017. "Gender's Critical Edge: Feminist Political Ecology, Postcolonial Intersectionality, and the Coupling of Race and Gender." In *Routledge Handbook of Gender and Environment*, edited by Sherilyn MacGregor, 146–158. London and New York: Routledge.
Murphy, Michelle. 2008. "Chemical Regimes of Living." *Environmental History* 13 (4): 695–703.
Rose, Deborah Bird, Thom van Dooren, Matthew Chrulew, Stuart Cooke, Matthew Kearnes and Emily O'Gorman. 2012. "Thinking through the Environment, Unsettling the Humanities." *Environmental Humanities* 1: 1–5.
Sandilands, Catriona. 2017. "Some 'F' words for the Environmental Humanities: Feralities, Feminisms, Futurities." In *The Routledge Companion to the Environmental Humanities*, edited by Ursula Heise, Jon Christensen and Michelle Niemann, 443–451. London and New York: Routledge.
Stengers, Isabelle. 2005. "Introductory Notes on an Ecology of Practices." *Cultural Studies Review* 11 (1): 183–196.
Tuana, Nancy. 2008. "Viscous Porosity: Witnessing Katrina." In *Material Feminisms*, edited by Stacy Alaimo and Susan Hekman, 188–213. Bloomington: Indiana University Press.

Part 1

Creatures

The Biopolitics of Making Kin

1 Mulberry Intimacies and the Sweetness of Kinship

Catriona Sandilands

Introduction: Mulberries, A Love Story

Until 2017, I never gave mulberries much thought. This situation is unusual, as I have been living intimately, thinking and writing with plants for a long time. If asked, I would have been able to hum the tune of "Here We Go 'Round the Mulberry Bush," but I had never knowingly tasted a mulberry fruit and certainly would not have said we were on close terms. Although a mulberry tree of some size inhabits the street-facing part of my front yard, I also had no real idea that I cohabited my West Toronto neighbourhood with rather a lot of its relatives. From a starting position of relative mulberry-blindness, that year I learned to pay attention to mulberry trees in two ways, more-or-less at the same time. In the first, I re-read Jeffrey Eugenides' novel *Middlesex* (2002) for a panel at the 2017 Association for the Study of Literature and Environment conference in Detroit, Michigan. I was looking for the plants in *Middlesex*, which is mostly set in Detroit, and there are mulberries all over it (both the novel and the city). In *Middlesex*, they are a major dramatic motif and their presence tendrils through the novel's stories of sex, gender, class, migration, racism and incest (to name a few, see Sandilands 2018). In the second, I took a field-naturalist course in Toronto's High Park: there are mulberries all over it, too. As the course leader outlined, these trees are also part of a drama, in which one kind of mulberry (*Morus alba*, or white mulberry) is considered an exotic-invasive, threatening the integrity of another kind of mulberry (*M. rubra*, or red mulberry) to the point the latter is listed as endangered under the Ontario Endangered Species Act. In High Park, we spent a lot of time learning how to distinguish the two species (fuzziness of leaves is the major indicator: you have to get pretty up-close and intimate with a mulberry to identify it), and it seemed that almost all the ones we examined were hybrids.

Especially with all this drama going on, I started to pay a lot more attention to mulberries. In fact, I fell completely in love with them. I learned to recognize their coyly variable leaf-shapes (heterophylly) and suddenly saw them everywhere: intensely pruned small, weeping varieties presiding over neat circular beds in highly manicured lawns; beautiful large trees shading

DOI: 10.4324/9781003023319-3

the streets of older neighbourhoods, often announcing their presence with extensive purple stains on the sidewalk; scraggly shrubs thriving at the edges of parks and, especially in Detroit, growing in glorious, entrepreneurial profusion in (now) vacant lots. I sought out their berries on local trees, where there were many, and in local shops, where there were none except dried ones. I sought them out in literary worlds beyond *Middlesex*, where they take many forms. In Ovid's *Metamorphoses* ([8] 2016) the ill-starred lovers Pyramus and Thisbe arrange to meet under a mulberry tree, and Thisbe's dying act is to curse its white fruit to turn the red of their blood. In Shakespeare's *A Midsummer Night's Dream* ([1595] 1895) they reappear via Quince's prologue in the play-within-a-play Pyramus and Thisbe; they also make their way into *Coriolanus*, where they are "humble" because delicate to handle. The image/idiom of "blue sea turned into mulberry fields" recurs in Chinese literature (e.g. in Ge Hong's third century CE *Shenxian Zhuan*, see Campany 2002) as an indication of epochal change; 1500 years later, Walt Whitman ([1855] 1904), also writing about change, includes a mulberry tree in his poem "This Compost" (in *Leaves of Grass*), which is a meditation on the nature of life and soul as an endless recycling of everyday matters and energies, including mulberry ones. And, in Hisham Matar's Booker-nominated *In the Country of Men* (2006) there is this gorgeous fragment from protagonist Suleiman, remembering his nine-year-old self in Tripoli:

> I decided that mulberries were the best fruit God had created and I began to imagine young, lively angels conspiring to plant a crop in the earth's soil after they heard that Adam, peace and blessings be upon him, and Eve, peace and blessings be upon her, were being sent down here to Earth as punishment. God knew, of course, He's the Allknowing, but he liked the idea and so let the angels carry out their plan. I plucked one off and it almost melted in my fingers. I threw it in my mouth and it dissolved, the small balls exploding like fireworks. I ate another, and another.
>
> (Matar 2006, 47)

Another, and another, and another. As I grew and savoured my intimacy with mulberries both local and literary – bloody, humble, sensuous, representative of life, change and love – it became increasingly clear to me that mulberry relationships are *often* very intimate. They are intimate in their bursting, luscious sweetness (Matar); in their association with blood (Ovid, Shakespeare); and in their multiple and ongoing involvements in economies of domestication (Ge Hong, Whitman, Eugenides, street and yard trees). Most obviously, of course, mulberries are intimately involved in the millennia-old industry of sericulture, in which they are the primary food source of the silkworm, *Bombyx mori*. Especially in sericulture, we see that these intimacies are not always sites of pleasure. For the domesticated silkworm, intimacy with mulberries and silk-harvesting humans not only leads to untimely death (the cocoons are generally boiled alive to clean and extract the finest silk thread) but has also, through thousands of years of selective breeding, rendered the adult moth incapable of flight and, perversely, insensitive to the *Morus* smell of the leaves on which it must lay its eggs (Coles 2019, 660).[1] For the white mulberry, however – the worm's favourite food – this intimacy has been largely responsible for the species' global distribution, from its origins in China to its now-ubiquity on all continents except Antarctica. As Peter Coles documents in his sumptuous plant biography *Mulberry* (2019), the globalization of sericulture and moriculture has been extensively bound up in imperialism, war, religious conflict and colonialism; both cultures have also supported artisanal as well as industrial forms of cultivation and production, and have extensively involved women's and children's labour both historically and in the present, including everything from empowering female entrepreneurship to institutionalizing child slavery.

It is now not uncommon to speak of "plant/human intimacies" as sites of multispecies enquiry. Julie Soleil Archambault's (2016) ethnography of gardeners in Inhambane, Mozambique, for example, suggests that such intimacies can embody the "transformative potential of everyday engagement with the material world" in opposition to neoliberal capital abstractions and the related "commodification of intimacy" (246, 247). I have – especially given the ways my moriphilic and other plant intimacies have shaped and enriched my inhabitation of Toronto – a great deal of sympathy for this argument. As Lauren Berlant, Kim TallBear and other feminist thinkers have emphasized in different ways, however, intimacies are also sites of regulation: who is allowed to be, forced to be and/or prohibited from being intimate with whom – and how – is a significant modality of biopolitical organization in settler-capitalist societies, and these operations of biopower clearly include plant and other multispecies actors. To practise restorative "kinship" with plants and others – as Robin Wall Kimmerer (2013), Deborah Bird Rose (2010), and Donna Haraway (2016) all suggest we do, urgently – thus must involve not only an embrace of plant-loving in general, but also a thoughtful questioning of the different kinds of intimacies in which we are already bound up with plants, and a careful exploration of other, perhaps more just and life-enriching intimate possibilities that may emerge in the process of enquiry. In this chapter, I will attempt some of this work of thoughtful questioning by focussing on mulberries, and specifically on the biopolitics of mulberry–human relations in Southern Ontario and on possibilities for their ethically sensuous reorientation in current conditions of intimate complexity. Specifically, I will tell two mulberry stories. In the first, I will consider the biopolitical organization of mulberry *species identity*, which is the central axis of the "invasion" drama involving *M. alba* and *M. rubra*. In this story, the reduction of plants to their genomic identities is part of the larger policing of "purity" that is central to the racist logic of settler colonization, that privileges the idea of a stable (and human-less) nature to be defended from change, and that prevents us from considering possibilities for *M. alba* intimacies outside their involvements in colonialism and invasiveness. In the second, I will consider the everyday and systemic regulation of mulberry *sexual expression,* which is linked to questions of utility; to a colonial distrust of fecundity, ferality, unpredictability and excess; and to the spatial and aesthetic impoverishment of multispecies intimacies and communities in cities like Toronto. Thinking about the biopolitical organization – and reorientation – of moricentric intimacies in these ways is a way of taking up possibilities for life and kinship in a manner that recognizes responsibility for the violences wrought by some institutions of intimacy, and that simultaneously creates space for ethical kin- and world-makings despite the uncomfortably intimate conditions in which we currently find ourselves.

Mulberry Intimacies I: Identity

Red mulberry is native to North America, where it is found throughout the central and eastern parts of the continent, especially in the southeast. Like all mulberries, *M. rubra* is a deciduous, fruit-bearing tree that grows quickly while young; it can reach a height of 65 feet and often lives 125 years (so: medium sized and not very old in tree terms).[2] Given its widespread distribution, sometimes dense population and prolific fruit production, it is not surprising that *M. rubra* – also *sachüissegóna òchia* (Onondaga), *oakhattim inschi* (Delaware), *mtekwa˙palwa* (Shawnee), *bihi* (Choctaw), *kė́* (Mvskoke/Creek) and *kuwa* (Cherokee), among other names – plays a major role in the culinary, medicinal and technological worlds of many Indigenous nations on the eastern side of the continent.[3] Indeed, in the southeast several nations celebrate a whole "mulberry month" – for example, *bihi hvshi* (Choctaw) and *Kē-Hvse* (Mvskoke), both roughly in May – out of respect and gratitude for their abundance. The fruit – carefully cultivated by some nations for millennia (see Taylor 2019; Treat 2011) – can be eaten fresh, dried, made into jam or jelly, and mashed into a variety of forms of cake and then reconstituted and cooked into a highly nutritious sauce for late winter days, high in Vitamin C, iron, K1, potassium and antioxidants. Small branches of red mulberry trees are pliable and can be made into bows and baskets; older wood is light, sturdy and good for construction; inner bark can be used to make cloth, without the intermediary of silkworms; and many parts of the tree create excellent dyes with colours depending on whether the roots, branches, bark or berries are used (Taylor 2019; Treat 2011, 2018). Different parts of the tree also contain bioactive compounds, and many nations have ongoing traditions of mulberry medicinal use as a laxative, emetic, cathartic, tonic, anthelmintic and/or hallucinogenic, and as a treatment for urinary problems, dysentery and ringworm.[4] *M. rubra* is at the far northern reach of its habitat in the Carolinian forests of Southern Ontario, which stretch southeast from what is now Toronto to Lake Erie (and beyond, southward). In Ojibwemowin, mulberries are *mitigwaabimin*;

in Mohawk, they are *shahieskó:wa*. In both nations, mulberries recede in importance relative to other berries such as strawberries, blueberries and bearberries, which are far more prevalent in these more northern territories; they are, however, important enough to have names.

White mulberry is, of course, a relative newcomer to North America, but it has been a globally cosmopolitan species for far longer. As Coles documents extensively, *M. alba* is originally from East Asia, where it has been "the stuff of legend" (2019, 62) for thousands of years, and was carried along the so-called Silk Road, slightly behind the moths, from China to Japan, India, the Middle East, Europe, and North Africa – where it met communities of black mulberry (*M. nigra*) – and beyond. Although less prized than that of *M. nigra*, the fruit of *M. alba* can be easily moved, dried and stored; perhaps more importantly, its leaves are the preferred food source of silkworms, even relative to other mulberry species, and so the trees have followed sericultural desires in their anthropogenic world travels rather than culinary, nutritional, medical or aesthetic ones. Such silk industrial aspirations brought *M. alba* to North America. Although European explorer/colonizers noted extensive, often carefully cultivated *M. rubra* groves in what are now South Carolina and Georgia in the sixteenth century, white mulberries were brought to the American colonies at about the same time with the specific intent of promoting a domestic silk industry. For example, in 1624 the legislature of Virginia required every (white, property-owning) male resident to plant at least four white mulberry trees for precisely this purpose (USDA Forest Service), and Benjamin Franklin, in the later eighteenth century, was instrumental in the planting of *M. alba* in cities like Philadelphia in order to make individual smallholdings more economically self-sufficient: "Mulberry Trees may be planted in Hedge Rows, or Walks or Avenues, or for Shade, near a House, where nothing else is wanted to grow" (Franklin 1769). The commercial "mulberry mania" that followed did not result in a lasting industry, at least partly because of a mulberry blight that decimated the planted trees and dashed settler-colonial sericultural hopes in the late nineteenth century[5]; however, white mulberry trees remain a fixture of North American cities and towns, and in the names of hundreds of Mulberry Streets whether or not there are any mulberries actually living on them.

As noted earlier, *M. alba* is, in North America, considered an invasive species, and this phenomenon has reached the most alarmist attention in Southern Ontario, where *M. rubra* has been listed as an endangered species partly because of *M. alba* incursions, and partly because of dramatic changes to *M. rubra*'s preferred habitats (*M. alba* is much less fussy). Although this scenario is hardly unique, the North American *Morus* story is complicated by *M. alba*'s "sociability," as Coles puts it (2019, 62): unlike its *M. nigra* cousin, white mulberry is strongly inclined to inter-breed with red mulberry, resulting in multiple hybrids wherever it meets them.[6] This – and *not* habitat competition – is the primary reason why *M. alba* is considered invasive in Southern Ontario: the species is increasingly "mixing" with *M. rubra*. As the Ontario Ministry of Natural Resources notes, there are fewer than 300 red mulberries remaining in Ontario, and "the greatest threat to Red Mulberry is cross-breeding with the non-native White Mulberry" (nd).[7] To spell out a key difference: it is not that individual red mulberry *plants* are suffering because of *M. alba* (in fact, they are thriving together as hybrids); it is that the purity of the native red mulberry *population* is called into question by its hybridization, and that they are an endangered *species* because the white ones are "mixing" with them. Despite the fact that *M. rubra/alba* hybrids are likely *increasing* their northern range and abundance along with urbanization and development (which *M. alba* enjoys and *M. rubra* clearly doesn't on its own), red mulberries are thus at-risk, and white mulberries are reviled for their reproductive promiscuity.

This understanding of the plant and its complex, international speciation – in the midst of a prevalent discourse of native/exotic, invasible/invasive, endangered/aggressive – is what gives me pause. Why was I taught to lament the hybridity of the tenacious little mulberries growing, fruiting and thriving in the middle of the highly disturbed and polluted landscape of High Park? Who says it is important that there be un-hybrid red mulberries in Southern Ontario, especially when there are a lot of them in the species' preferred, more southerly latitudes? Why, when mulberries as an exuberant and fruity whole are clearly flourishing at the very northern extremes of *M. rubra* territory, is their apparently voluntary and enthusiastic mixing considered a botanical disaster, especially given that air pollution and habitat destruction affect *M. rubra* much more than *M. alba*? Several sources make the stakes clear. Most prominently, Parks Canada notes that "the large number of White Mulberry trees and hybrids across the landscape, and the genetic makeup of the hybrids, suggest that the Red Mulberry is being *genetically assimilated* by White Mulberry" (2011 4.1; my emphasis). This conclusion is based on an earlier study by Burgess et al, which observes that, although *all* the hybrids they sampled contain the genetic materials of *both* species, there was a "relative bias" of *M. alba* nuclear markers in 67 per cent of them, a situation of "asymmetrical introgression" (2005, 3471) most likely caused by the larger pollen-producing population of *M. alba* in the study areas (could one conclude that, in a more rubra-abundant location, the asymmetry might be reversed?). Even from the narrow perspective of *M. rubra* genetic futurity, hybridization with the more cold-tolerant, atmospherically hardy, and disturbance-forgiving white mulberry seems an effective adaptive strategy more than it does an indicator of endangerment; however, what is clearly at issue for Parks Canada is neither the health of the plant nor its futurity, but the maintenance of a "pure" population of the species.

The logic of "genetic assimilation" is important in this situation for several reasons. First, it insists that the identity and importance of the plant are, first and foremost, a matter of genetics and not, say, of the plant's relationship to human or larger ecological communities. Although hybrid mulberries are doing well, including the continued, prolific production of berries that are important to many bird and animal species (including people), the insistence on the primacy of genetic purity in a larger narrative of species decline says a lot about the ecological and cultural priorities involved in the story. Mulberries' continuing gifts to multispecies relationships are overshadowed by a desire for species distinction. Although the two relations are by no means the same, this emphasis on mulberry purity looks a great deal like what has been named the "vanishing Indigene" trope in which colonizers expressed nostalgia for the loss of "pure" Indigenous peoples as a way of drawing attention away from continued, genocidal treatment of Indigenous people who persist in the midst of ongoing settler-colonial relations (TallBear 2013). This narrative is a way of romanticizing a static, Edenic past in the service of making the present condition "fallen" and therefore profane and unworthy of care (Shotwell 2016). Further, it is a way of drawing attention away from the structural relations that create the conditions for the loss in the first place: in the case of Indigenous peoples, ongoing, violent efforts by the settler-colonial state to impoverish and disempower Indigenous nations and sovereign land and kinship relations, and in the case of native species, ongoing work to destroy and pollute lands, habitats, and plant and animal relational ecologies. Both the Canadian and US states have used arbitrary measurements such as "blood quantum" to regulate and impoverish Indigenous family relations and land rights (TallBear emphasizes ongoing complexity); in the case of hybrid mulberries, branding them less-than-pure has the effect of making them infinitely disposable in the service of maintaining a conservative and rigid roster of "native" flora and fauna. As Alexis Shotwell writes, "to be against purity," in this sense, is "not to be for pollution, harm, sickness, or premature death. It is to be against the rhetorical or conceptual attempt to delineate and delimit the world into something separable, disentangled, and homogeneous" (2016 15). For mulberries, it is also about being aware of the specific work that the idea of "endangerment through hybridization" brings to the larger constellation of racialized, colonial practices through which bodies and multispecies relationships are figured and managed by the settler-colonial state.

In her powerful analysis of recent trends toward understanding Indigenous people and peoples in terms of genomic articulations of identity and ancestry, TallBear strongly criticizes biocolonial reductions of indigeneity to molecular sequencing in human genome diversity research (and its political applications). She writes that

> genomic indigeneity is an articulation that focuses on biological descent and relations between groups across time and space. Within this articulation, an indigenous group becomes a biological-based or population-based category [...]. What unites indigenous peoples globally, from a genomic perspective, is not opposition to colonialism or autochthonous cosmology, but relatively straightforward genetic descent from founder populations on particular continents.
>
> (TallBear 2013, 516)

Although many Indigenous people themselves "also privilege biological connection to ancestors (alongside connection to land)," they have "evolved a more multifaceted definition of 'indigenous' that entangles political self-determination and mutual networking for survival" (510). Thinking with TallBear, it seems pretty clear that Indigenous people likely have something more complex and nuanced to say about mulberry identity as well: perhaps *Morus* hybridization could also be considered a form of "mutual networking for survival," especially given the strong relationship between the dispossession of Indigenous lands and the destruction of red mulberry habitats in Southern Ontario, and also the need to consider, in these conditions, how *survival* might need to involve *adaptation*, as it has done for Indigenous peoples for millennia (Reo et al. 2017).

Further, this understanding of *M. alba* "invasion" in Southern Ontario is not necessarily in accord with Anishinaabe or Haudenosaunee understandings of invasive species. Although Indigenous nations, communities and individuals lead multiple forms of plant care and habitat restoration and there is no singular view of invasive species among them (Reo et al. 2017; Indigenous Land Stewardship Circle nd; Corntassel 2020), one thing that is clearly important is what Reo and Ogden (2018) call the "invasive land ethic" in which all species changes are occurring in settler-colonized landscapes, including both settler-capitalist development and aggressively managerial forms of conservation. I am not an Indigenous person with a multi-generational tradition of intimacy with red mulberries. However, I have to question if changing species identity is anywhere near as important as the changed relationships that surround current mulberry transformations. As the Indigenous Stewardship Circle counsels in relation to programmes of invasive species eradication, "stigmatizing our Relatives in this way would make it impossible to learn what it is that they are here to teach us" (nd). Perhaps the lesson is that we need to *listen* to intra-mulberry intimacies, and what they are telling us about changes to the land, rather than demonizing them.

Again, I am *in no way* arguing that the regulation of Indigenous peoples' intimacies and relations is equivalent to the regulation of native plant intimacies and relations. Quite the opposite: I think it is deeply troubling that Parks Canada, an apparatus of the Canadian state, is so concerned about indigenous plants in the midst of the same state's ongoing dispossession of Indigenous communities, including their intimate *relations* with native plants as part of multispecies family networks, and also including the "vicious restraint" (TallBear 2018, 147) of Indigenous kinship relations more widely. Settler movements for the preservation of native plant species *very often* proceed without attention to the larger Indigenous kinship networks in which plants and people are intertwined: the plants can be "protected", in a settler-colonial understanding of the land, without addressing questions of Indigenous land rights and relations, because of settler-colonial fantasies that the original state of the land is an unpopulated "wilderness", and not a matter of, for example, ongoing plant/people collaborations that involve movement, change, and co-adaptation. This colonial logic of plant indigeneity is a way of settlers keeping an idea of wilderness, and continuing to displace the people (and their relations to the plants) with whom the land is intimate: it is not possible to engage in "botanical decolonization" (Mastnak et al. 2014) without attending to the people with whom the plants are so deeply enmeshed (Taylor 2019; Treat 2011, 2018), and without attending to what the plants are teaching *all* of their human cohabitants.

More specifically, I would argue that the "mulberry problem" in Southern Ontario is not one of *M. alba* incursions into *M. rubra* species "integrity"; it is a problem of *wastelanding*, in which the robust spaces of ongoing people/plant relationships have been decimated to make way for settler-capitalist developments that destroy any but the most instrumental and profitable ones (see TallBear 2018). Other major threats to red mulberries listed in the Parks Canada report are habitat fragmentation and pollution: why are they not at the very top of the endangerment cause list, especially when the combination of the two leads to further, *M. rubra*-specific problems of stress and disease? The pursuit of pure *M. rubra* is, in this context, a deeply problematic attempt to preserve a dubious, colonial understanding of a plant's species futurity without consideration of the people/plant futures of which the species is part. The question of *M. rubra* endangerment can, then, be seen in a different light: perhaps it is more important to think about the ongoing possibilities for relationship between people and mulberries than it is to insist on a particular, genomic species identity? Perhaps it is more important to look at the possible ongoingness of human/animal/mulberry intimacies than it is to try to regulate white/red mulberry intimacies, as if the plants – and not the settler-capitalist decimation of *M. rubra* relational landscapes – are the cause of the problem?

Mulberry Intimacies II: Excess

Mulberries are abundant in cities in the eastern part of North America. Some *M. rubra* were there before the cities were. Others, likely *M. alba*, remain because of their past incorporations in sericultural desires; many of their offspring – including multiple hybrids – also persist in the ruined landscapes of the Rust Belt such as Detroit, in which their gregarious and tenacious habits have found purchase in neglect, to the point that mulberries are now an unofficial emblem of the city. Mulberries have also, however, found conditions in which to thrive in urban landscapes because they are excellent street trees: *M. alba* grows quickly, is resistant to air pollution and provides ample shade because of its dense canopy once matured. In my Toronto neighbourhood, there are many planted mulberries of different kinds, sizes and ages, including the one in my front yard: their leaves turn a very attractive shade of yellow in the autumn, and their presence is – once you start to notice them – a vibrant addition to the streetscape.

The most noticeable mulberry trees in my neighbourhood are also prolific producers of berries.[8] A City of Toronto website includes nearly 2000 mulberries on its interactive City Fruit Tree Map (Weatherburn and Davis 2017). The local community gleaning organization Not Far From the Tree offers tips for harvesting these urban mulberries (which it suggests are mostly hybrids, but which are likely often just *M. alba*): the "shake and catch" method, respecting both the trees' height and their month-long fruiting period, suggests that you

> grab a few friends and a large tarp or sheet. Have four people hold each corner of the tarp, while one person climbs the tree or reaches up to shake the branches. The mulberries will rain down on the tarp (and onto people's heads) [...] avoid wearing fancy clothes.
>
> (2019)

Many local walkers and foragers welcome mulberry abundance, which reaches its peak in early July in Toronto. However, mulberries definitely make a mess on the pavement, both because they are extraordinarily prolific in their fruiting and because they are very popular with urban birds such as cardinals, robins, crows, jays, orioles and mockingbirds (and also foxes, raccoons, skunks, squirrels and opossums, even if their mulberry-coloured excrement is a bit less randomly distributed).

Because of mulberry fecundity and, especially mess, many cities have heavily favoured male mulberries (no berries) over female ones in their organized, street planting schemes: all the benefits of a hardy, pollution-resistant, fast-growing, attractive, mid-sized tree, but with none of the sidewalk stains or Jackson Pollock purple poop. However, as several US cities have acknowledged with legislation banning the planting of mulberries of any sort (Tucson, El Paso, Albuquerque, Las Vegas),[9] a preponderance of male mulberries creates a considerable problem. Specifically, male mulberries are prolific producers of *pollen*, and "exposure to pollen from white mulberry has been reported to cause asthma, allergic rhinitis, allergic conjunctivitis, and symptoms of contact urticaria" in studies dating back to 1971 (Papia et al. 2020, 2).[10] Of course, many trees – especially wind-pollinated species like mulberries – produce allergic reactions in people because their pollen is light and broadcast widely. However, male *M. alba* are quite exceptional. Taylor et al found that, in experimental conditions, their inflorescences ejected pollen at a speed of *Mach 0.7.* As they describe:

> Exposure to dry air initially resulted in the gradual movement of the stamen. This caused fine threads to tear at the stomium, ensuring dehiscence of the anther, and subsequently enabled the anther to slip off a remaining pistillode. The sudden release of stored elastic energy in the spring-like filament drove the stamen to straighten in less than 25

> μs, and propelled the pollen grains to velocities in excess of half the speed of sound. *This is the fastest motion yet observed in biology, and approaches the theoretical physical limits for movements in plants.*
>
> (Taylor et al. 2006, 19; my emphasis)

Thomas Ogren is a self-proclaimed "horticultural epidemiologist" and creator of the world's first numerical plant allergy scale (OPALS); he has written extensively about the pollen proliferation of mulberries and other allergenic street trees. As he wrote in a recent blog in *Scientific American*: "decades of *botanical sexism* in urban landscapes have contributed" (2015; my emphasis) to street canopies that, as a result of preferential planting of "litter-free" male trees for decades – including industrial-scale, clonal propagation of male trees from other male trees – are far more allergenic than those of cities prior to the 1970s (when Dutch Elm Disease hit and the dead trees were replaced by fast growing, pollution-hardy species like Norway maple and mulberry). He strongly advocates the planting of female trees, which do not produce pollen; in fact, they attract, trap and metabolize it (which is, of course, the reproductive point of the pollen in the first place). Moreover, he advocates both the removal and the alteration of male trees: "those extraordinarily allergenic mulberry trees could be made less irritating to allergy sufferers if they were given a *sex change*, that is, top-grafted with weeping mulberry, which is female and pollen-free" (2010; my emphasis). Although his faith in weeping mulberry's singular sex expression may be misplaced (more on this below), he is drawing on centuries of arboriculture and moriculture in his understanding that grafting (one method of clonal reproduction) is a very effective way of shaping trees to human desires: less preferred rootstock can be made productive (or not) by the skilful attachment of the scion of a sought-after variety or, in this case, sex, and the repeated cutting away of anything sprouting below the graft. A "male" tree can thus become, functionally, "female," thus eliminating (or at least mediating) a different mulberry excess, in this case its abundant and impressively ballistic pollen ejaculations.[11]

The irony of this situation should not be lost: male mulberries were preferentially and systematically planted in urban environments on a large scale to avoid female, fruiting mulberry mess, and now precisely these female characteristics are being called upon to mitigate a new, allergenic "mess." Although human control and shaping of plant sexuality is a practice with an ancient and varied history, including everything from strategic planting to meticulous hybridization, there is thus something quite specific about the sexual regulation of mulberries to conform to utilitarian and aesthetic urban desires that merits attention. What is it about mulberries that, apparently, demands such close sexual scrutiny and selection? Why are mulberries – especially given their generous and prolific production – not highly prized alongside, say, apples, cherries, plums and pears in urban landscapes like Toronto?

One clue about the apparent problem of mulberry "excess" lies in their fruit and fruiting habits. Specifically, mulberries ripen over a period of several weeks and, further, the berries on older trees are often not within easy human reach. The "shake and catch" method of harvest is all well and good, but the mulberries cannot all be caught in one shake because they do not all ripen at the same time: great for birds and raccoons and mulberry futurities, and also potentially great for people interested in an accessible food source for a relatively long period of time, but not so great for potential mulberry users who might rely on a stable, harvestable crop for longer-term consumption and processing. Further, mulberries do not keep or travel well; one very seldom sees fresh mulberries for sale, even in ultra-local farmers' markets, as a result. Although *M. alba* dries well whole, which is why I can find it at some local stores (*M. nigra* does not and needs to be processed quickly, and *M. rubra* is also usually processed before drying in Indigenous food traditions), it is clear that mulberries are not an exceptionally good fruit for commodity purposes. They are best enjoyed just off the tree, and with relish, as in Matar's sensuous account of the berries "exploding like fireworks" in his mouth (2006, 47); as in the climactic scene in Eugenides' *Middlesex*, in which the protagonist, Cal, sits at a second floor window plucking mulberries of the tree, "fat and juicy […] they stained my hands" (2002, 396); and as, perhaps most especially, in lifeworlds that genuinely appreciate the important seasonality of mulberry fruit over their capacity for storage, transport or sale (e.g. Taylor 2019).[12]

So: fruiting mulberries tend to thwart an *accumulative* logic: their generosity tends to be in-the-moment rather than for profit. James Treat (2011) underlines this difference in his comparison of Mvskoke cultivations of *M. rubra* for food and other uses, and colonial plantations of *M. alba* to feed silk production: "As was so often the case, immigrants saw profit where Indians had seen only provision". Further, as Kimmerer (2020) writes in a powerful essay on serviceberries, *Amelanchier arborea* – or *Bozakmin* in Potawatomi, the "min" connecting to them to mulberries through a root that also founds the word "gift" even if the two plants are not related in Eurowestern taxonomies – there is a large difference between appreciating their abundant fruit as gifts to be shared in multispecies community, and understanding their largesse as an opportunity for industry. As she writes, "when we speak of these not as things or products or commodities, but as gifts, the whole relationship changes. I can't help but gaze at them, cupped like jewels in my hand, and breathe out my gratitude (2020)". Serviceberries, like mulberries, give generously, in a way that invites ecological reciprocity (e.g., in their attractiveness to birds, whose digestive processes scarify the seeds to promote germination via their abundant purple poop). Serviceberries – and, I think, mulberries – also hold the potential to inspire and model reciprocity in human communities. Kimmerer writes about Indigenous gift economies, and of the crucial difference between an understanding of seasonally berrying plants as bearers of abundance, for sharing, and a view of them as problems to be solved for accumulation:

> The [serviceberry] gift is multiplied with every giving, until it returns so rich and sweet that it burbles forth as the birdsong that wakes me in the morning. If the abundance had been hoarded, if Juneberries acted solely for their own benefit, the forest would be diminished.
>
> (Kimmerer 2020)

Think of the difference, then, between understanding fruiting urban mulberries as a mess, and viewing them as a reminder of webs of plant-animal reciprocity that persist, and that need to be tended, even in the deeply compromised relations of large cities like Toronto: rather than focussing on planting or creating female mulberries as a gesture of "controlling" the ejaculations of over-planted male trees, what would it mean to think about creating urban moriscapes that invite an understanding of mulberries as part of a movement toward renewed reciprocity, and toward relationship rather than control?

Further, mulberries often *resist* anthropocentric desires for mulberry sexuality: as most anyone who has tried to grow them for fruit can tell you, their sexual expression is not fixed.[13] Not all that unusually in the plant world, some mulberry trees express as dioecious (presenting either male or female inflorescences) and some as monoecious (including both); moreover, any given tree can change its sexual expression depending on factors including the presence of other trees, environmental conditions and even individual trauma such as pruning (Tikader et al. 1995). In the first year of my mulberry love affair, I visited a Toronto big-box home store in search of information about the kinds of mulberries that are currently marketed for landscaping and home gardening: *every single one* of the small, weeping trees on sale as specimens of a "non-fruiting variety" bore prolific, luscious berries that were, not surprisingly, making a mess of the display in multiple ways. In short, mulberries are not necessarily inclined to perform the sex/gender expectations of either urban planners (and epidemiologists) or fruit growers: their sexual expression is part of a mulberry logic that does not always conform to what people want.

People who cultivate and interact with mulberries intimately – both for fruit and for other needs – have thus paid intimate attention to the sex lives of the plants for a very long time. Although some mulberries can present as only female and still set fruit, the majority of them thrive in situations in which there is more sexual diversity in the moricentric landscape: both trees that exhibit female flowers to set fruit, and ones that exhibit male flowers to provide enabling pollen, on the same or different trees. There is, however, a large difference between human cultivation of mulberries in a way that may shape, but still contributes to, *their* preferred sexual relations, and human control of mulberries in a way that intentionally *obstructs* these relations for industrial or planning purposes, as is the case with a monoculture of male mulberries as street trees or with mass grafting to make them female. Mulberries themselves have the capacity to thwart these industrial, anthropogenic desires for "litter free" – or, indeed, for hypoallergenic – trees by changing sex of their own accord.

Why are we not listening to them? Rather than consider mulberries as antagonistic sexual subjects, whose desires and intimacies need to be forced into monocultures in the service of specific urban, industrial needs, what might it look like to work *with* them to create conditions for mutual flourishing? How do the trees, with their exuberantly excessive berries and/or their extraordinarily projectile pollen, suggest ways for people to cohabit, intimately, with mulberries that are not primarily written in a language of botanical sex control? Why not embrace the generosity and create conditions in which the pollen is not so much an excess to be controlled as it is part of a cycle to be cherished? I have thought about doing some grafting with the tree fronting the street: although, as one of my non-binary friends has pointed out, the tree might not have invited this experiment, perhaps the addition of some berry-producing scions to the top of the tree – without pruning away the male inflorescences at the bottom – might be a reasonable way of inviting diversities of mulberry sexual expression into the streetscape?

Conclusion: Sweet Mulberry Kinship

In her poem "Becoming Seventy," Mvskoke poet Joy Harjo writes that

> we all have mulberry trees in the memory yard. They hold the place for skinned knees earned by small braveries, cousins you love who are gone, a father cutting a/watermelon in the summer on the porch, and a mother so in love that her heart breaks – it will never be the same, yet all memory bends to fit.
>
> (Harjo 2016)

The idea that mulberry trees might be holders and keepers of memory in a world that will "never be the same" is an apt place to conclude these explorations. In addition, according to Treat *Kē-Hvse* is an annual reminder of how the people (including nonhuman people) came together to overcome the trauma of creation: After a winter of "bitter memory and lesson [...] at the time of the Mulberry Blossom, all shall come together again and seek to renew these teachings" (2011). Mulberries are both holders of memory in the midst of change, then, and also reminders of the importance of multispecies "comings together" in the work of survival and renewal in these traumatic times.

I am not ashamed of the promiscuity of my mulberry relationships. I love their presence in my inheritance of Shakespeare; in their invitation to experience Tripoli and Detroit through sensuous connection with their fruit; in the complicated stories they have told me about seasons, cakes, bows, berries, dyes and silk; in the lessons they have taught me about changes to the land I live in; and in the important things they and their Indigenous interlocutors have reminded me about histories, relationships and obligations in the midst of my complicity in settler-colonialism. Most particularly, my mulberry love-affair has led me to think about intimacy in complex, moricentric ways: how the sex and species intimacies of mulberry trees are bound up with global, colonial, multispecies biopolitics, including the intersecting regulations and productions of both human and nonhuman intimacies; and how these specific moricentric relationships are involved in larger issues of living, and loving, in the current conditions of what Anna Tsing (2017) has appropriately called "capitalist ruins".

As TallBear writes, "recognizing possibilities of other kinds of intimacies – not focused on biological reproduction and making population, but on caretaking precious kin *that come to us in diverse ways* – is an important step to unsettling settler sex and family" (2018, 154; my emphasis). Understanding our relationships with mulberries *as* intimacies invites us both to think carefully about *how* they have "come to us," and to relate to them in ways that are based in practices of respect and mutual flourishing rather than eradication and control. As I have explored in this chapter, mulberries have a lot to tell us about histories, relations, desires and futurities: ours, theirs and all together. In addition, they invite possibilities for a concrete practice of close and loving cohabitation: becoming intimate with mulberries demands that we ask how they invite us to join into *their* webs of family and memory at the same time as we come to value them as *our* generous kin. In addition to listening respectfully to, and learning from, long traditions of mulberry and other multispecies kinship that emerge from Indigenous philosophies and practices of reciprocity and relationship, this concrete practice involves – for settler-colonial moriphiles like me – letting go of the desire to *manage* mulberry identities (understood as populations, resources and problems) in favour of working *with* them to contribute to conditions of shared nourishment. As Kimmerer writes, "both the honor of giving and the humility of receiving are necessary halves of the equation" (2013, 381); indeed, perhaps once more people start to care *for* mulberries, more will also start to notice, and relish in, their abundant gifts.

Appreciating the sweetness of mulberry kinships should not be indiscriminate and does not mean simply leaving current mulberry worlds to continue as they are: remember that *cultivation* is a long, diverse and multiply storied co-creation between mulberry and human (and other) futurities, and that part of the problem in Southern Ontario may be that attentive cultivation is precisely what is lacking. Just because *M. alba/rubra* hybrids are doing pretty well in devastated landscapes and streetscapes does not mean that mulberry lovers should not try to protect, create and nurture habitats in which *M. rubra* relationships have a greater chance of thriving. Just because it would be hubristic folly to try (and likely fail) to forcibly change the sex expression of thousands of male-presenting trees in some grand reversal of earlier urban landscaping mistakes does not mean that it might not be worth trying to work with the tree in my front yard, over time, to see if we couldn't come up with a satisfactory compromise about its inflorescences, perhaps including the addition of another mulberry to the area. What a treat it would be to be able to pluck a handful of ripe berries for a July breakfast in exchange for some nice, potassium-rich organic fertilizer and regular watering, something that many street trees are routinely denied. What a treat it would be to know that the mulberry in my front yard – like the serviceberry in my back yard – is contributing to multiple bird and mammal worlds as well as my own? There is a large difference between mulberries as objects of intimate intervention and as partners in intimate cohabitation. It is high time, as Coles suggests, to "unshackle" mulberries from their "5,000 years of bondage" (2019, 2682) in sericulture and other imperialist and capitalist desires and use our closeness to them as a starting-point for more mutually enriching tendings.

Notes

1 Coles adds that "recent genetic analysis has shown that the relationship between mulberry and silkworm is even more intimate than was ever imagined: fragments of mulberry plant genetic material have somehow been incorporated in to the silkworm itself" (2019, 669).
2 There are three main species of mulberry in the family *Moraceae* (white, red and black) and multiple cultivars. Over 200 separate species are identified in taxonomy but, as the chapter will discuss, mulberry identity is sometimes tricky. This chapter will only discuss *M. nigra* in passing; it is, however, widely reputed to have the most delicious fruit and, although it is not preferred by silkworms, it is often the choice of fruit growers.
3 There is relatively little published material about red mulberry. The source on which most publications and databases rely is Daniel F. Austin's monumental *Florida Ethnobotany* (2004); his sources are, in turn, largely earlier settler-colonial botanists and ethologists. Other, more primary sources include a recent book published by the Choctaw Nation (excerpted in Taylor 2019), an essay by James Treat in the *Mvskoke Nation News* (2011), and a found poem compiled by Treat from interviews with Elders, "They Used Mulberry Tree" (2018).
4 As both Austin and Coles demonstrate, there are many more written sources, as well as recent peer-reviewed studies, on the medical uses of *M. alba*, which has been found to be especially useful for the treatment of Type 2 diabetes (including but not only in Traditional Chinese Medicine). It is likely, but not demonstrated, that all three species have similar bioactive properties; this question is important to consider in relation to inter-*Morus* hybridity and Indigenous futures.
5 It is worth noting that an utopian community of abolitionists including Sojourner Truth bought a Connecticut property specifically to produce silk as an alternative to slave-dependent cotton ("Connecticut's Great Mulberry Mania").
6 Given that a "species" is typically understood as the largest group of organisms in which any two individuals can sexually re/produce fertile offspring – which red/white mulberry hybrids clearly are – it is questionable whether white and red mulberries are really separate species, and/or whether the organismic desire of these plants themselves is simply toward hybridity rather than separateness,

calling into question the whole idea of "species" in this case (and in the case of many other plants such as mustards).

7 In order, Parks Canada lists the following as the most pressing threats to *M. rubra* survival in Southern Ontario: hybridization with white mulberry; habitat loss and fragmentation; nesting double-crested cormorants (on whom I have also written a paper: Sandilands 2016); disease and causative stress factors (including pollution); and other threats and exotic species, including herbivory by animals. The Report is not clear about the rationale behind the ranking; however, it seems clear that the disruption of the Carolinian ecosystem in general is the major culprit behind *M. rubra* decline, and it is interesting that *M. alba* gets the "blame".

8 Mulberry fruits are, technically, not berries: they are "multiple," "collective" or "aggregate" fruits, meaning that each mulberry flower in an inflorescence (group of flowers) produces an individual fruit that ripens into a cluster, as opposed to a true berry, which results from a single flower.

9 Also Islamabad and other cities in Pakistan, although the species in question there is paper mulberry, *Broussonetia papyrifera.*

10 In a review of the literature, Papia et al. (2020) conclude that mulberry pollen is not only allergenic on its own but also in combination with that of other common urban tree species such as birch.

11 The whole question of "sexing" plants calls into question the applicability of understandings of stable sex (and even gender) that are largely shaped by Eurowestern assumptions about sex and sex expression associated with, mostly, mammalian life forms. Plant "sex" is often simply a question of which inflorescence(s) are present on a given plant in a given blossoming season: the assumption that sex (and gender) is part of an ongoing organismic identity is clearly not one modelled on plants, many of which change inflorescence in response to environmental and other conditions.

12 There is also, of course, a seasonality to mulberries' many non-fruit uses: according to Taylor (2019), the best time to strip bark from mulberries for textile production is also in *Bihi Hvshi.*

13 They are also heterozygous (like people), meaning that their offspring are not true-to-type, so mulberry cultivation frequently involves clonal reproduction as well as growing from seed.

References

Archambault, Julie Soleil. 2016. "Taking Love Seriously in Human-Plant Relations in Mozambique: Toward an Anthropology of Affective Encounters." *Cultural Anthropology* 31 (2): 244–271.

Austin, Daniel F. 2004. *Florida Ethnobotany.* Boca Raton, FL: CRC Press.

Burgess, K.S., M. Morgan, L. Deverno and B.C. Husband. 2005. "Asymmetrical Introgression Between Two *Morus* Species that Differ in Abundance." *Molecular Ecology* 14: 3471–3483.

Campany, Robert Ford. 2002. *To Live as Long as Heaven and Earth: Ge Hong's Traditions of Divine Transcendents.* Berkeley, CA: University of California Press.

Coles, Peter. 2019. *Mulberry.* London: Reaktion Books. (Kindle edition)

"Connecticut's Great Mulberry Mania of the 1830s." https://www.newenglandhistoricalsociety.com/connecticuts-great-mulberry-mania-1830s/.

Corntassel, Jeff. 2020. "Restorying Indigenous Landscapes: Community Regeneration and Resurgence." In *Plants, People, and Places: The Role of Ethnobotany and Ethnoecology in Indigenous Peoples' Land Rights in Canada and Beyond*, edited by Nancy J. Turner, 350–365. Montreal: McGill-Queen's University Press.

Eugenides, Jeffrey. 2002. *Middlesex*. New York: Farrar, Strauss and Giroux.

Franklin, Benjamin. 1769. "Letter from Benjamin Franklin to Cadwalader Evans, 7 September." https://founders.archives.gov/documents/Franklin/01-16-02-0113.

Haraway, Donna. 2016. *Staying with the Trouble: Making Kin in the Chthulucene*. Durham, NC: Duke University Press.

Harjo, Joy. 2016. "Becoming Seventy." Poetry Foundation. https://www.poetryfoundation.org/poetrymagazine/poems/143935/becoming-seventy.

Indigenous Land Stewardship Circle. nd. "What Is Indigenous Land Stewardship?" https://indigenouslandstewardshipto.wordpress.com/indigenous-stewardship-vs-chemical-management/.

Kimmerer, Robin Wall. 2013. *Braiding Sweetgrass: Indigenous Wisdom, Scientific Knowledge, and the Teachings of Plants*. Minneapolis: Milkweed Editions.

Kimmerer, Robin Wall. 2020. "The Serviceberry: An Economy of Abundance." *Emergence Magazine*, December 10. https://emergencemagazine.org/story/the-serviceberry/.

Mastnak, Tomaz, Julia Elyachar and Tom Boellstorff. 2014. "Botanical Decolonization: Rethinking Native Plants." *Environment and Planning D* 32: 363–380.

Matar, Hisham. 2006. *In the Country of Men*. New York: Penguin.

Ministry of Natural Resources (Ontario). nd. "Red Mulberry/*Morus rubra*."

Not Far From the Tree. 2019. "Mulberries." https://notfarfromthetree.org/fruit/mulberries/

Ogren, Thomas Leo. 2010. "Allergy Free New York." *New York Times*, April 5. https://www.nytimes.com/2010/04/06/opinion/06ogren.html.

Ogren, Thomas Leo. 2015. "Botanical Sexism Cultivates Home-Grown Allergies." *Scientific American*, April 19. https://blogs.scientificamerican.com/guest-blog/botanical-sexism-cultivates-home-grown-allergies/.

Ovid. (8) 2016. *Metamorphoses*. Translated by David Raeburn. London: Penguin Classics.

Papia, F., C. Incorvaia, L. Genovese, S. Gangemi, and P. L. Minciullo. 2020. "Allergic Reactions to Genus *Morus* Plants: A Review." *Clinical and Molecular Allergy* 18 (1): 1–5.

Parks Canada Agency. 2011. *Recovery Strategy for the Red Mulberry (*Morus rubra*) in Canada*. Species at Risk Act Recovery Strategy Series. Ottawa, ON.

Reo, Nicholas J. and Laura A. Ogden. 2018. "Anishnaabe Aki: An Indigenous Perspective on the Global Threat of Invasive Species." *Sustainability Science* 13: 1443–1452.

Reo, Nicholas J., Kyle Whyte, Darren Ranco, Jodi Brandt, Emily Blackmer and Braden Elliott. 2017. "Invasive Species, Indigenous Stewards, and Vulnerability Discourse." *American Indian Quarterly* 41 (3): 201–223.

Rose, Deborah Bird. 2010. "Flying Fox: Kin, Keystone, Kontaminant." *Manoa* 22 (2): 175–190.

Sandilands, Catriona. 2016. "Some 'F' Words for the Environmental Humanities: Feralities, Feminisms, Futurities." In *The Routledge Companion to the Environmental Humanities*, edited by Ursula Heise, Jon Christensen and Michelle Niemann, 427–435. London: Routledge.

Sandilands, Catriona. 2018. "Mulberridlesex." *The Goose* 17 (1), Article 56.

Shakespeare, William. (1595) 1895. *The Works of William Shakespeare with a Memoir, Glossary, Etc*. London: Frederick Warne & Co.

Shotwell, Alexis. 2016. *Against Purity: Living Ethically in Compromised Times.* Minneapolis: University of Minnesota Press.

TallBear, Kim. 2013. "Genomic Articulations of Indigeneity." *Social Studies of Science* 43 (4): 509–533.

TallBear, Kim. 2018. "Making Love and Relations beyond Settler Sex and Family." In *Making Kin Not Population*, edited by Adele E. Clarke and Donna Haraway, 145–164. Chicago, IL: Prickly Paradigm Press.

Taylor, Ian. 2019. "Bihi Hvshi: Mulberry Month Corresponds with Spring," excerpted from *Choctaw Food: Remembering the Land, Rekindling Ancient Knowledge.* Choctaw Nation: Choctaw Nation Education Special Projects. http://www.choctawnation.com/sites/default/files/Iti%20Fabvssa_May2019.pdf.

Taylor, Philip E., Gwyneth Card, James House, Michael H. Dickinson and Richard C. Flagan. 2006. "High-speed Pollen Release in the White Mulberry Tree, *Morus alba L.*" *Sexual Plant Reproduction* 19 (1): 19–24. (incl. erratum)

Tikader, A., K. Vijayan, M.K. Raghunath, S.P. Chakroborti, B.N. Roy and T. Pavankumar. 1995. "Studies on Sexual Variation in Mulberry (*Morus* spp.)." *Euphytica* 84: 115–120.

Treat, James. 2011. "Kē-Hvse, "Mulberry Month." *Mvskoke Nation News,* May 1. https://mvskokecountry.online/tag/hvse/.

Treat, James. 2018. "They Used Mulberry Tree." *Westview* 34 (1), Article 12. https://dc.swosu.edu/westview/vol34/iss1/12.

Tsing, Anna Loewenhaupt. 2017. *The Mushroom at the End of the World: On the Possibility of Life in Capitalist Ruins.* Princeton, NJ: Princeton University Press.

USDA Forest Service, "*Morus alba,*" Fire Effects Information System (FEIS). https://www.fs.fed.us/database/feis/plants/tree/moralb/all.html#:~:text=White%20mulberry%20is%20native%20to, by%20%5B18%2C33%5D.

Weatherburn, Tom and William Davis. 2017. "The Fruit Trees of Toronto." City of Toronto. http://www.mapto.ca/maps/the-fruit-trees-of-toronto#.

Whitman, Walt. (1855) 1904. *Leaves of Grass.* Boston: Small, Maynard & Co.

2 The Vegetal Subjects of Feminist Speculative Fiction

Natania Meeker

The past ten years or so have witnessed the rise of critical plant studies[1] in the humanities – a phenomenon sometimes referred to as the plant turn. Often in dialogue with scientific findings on plant functions, contemporary plant theorists consider vegetal life as its own autonomous mode of being, endowed with specific energies and forces and capable of interacting with humans on its own terms. Critical plant studies blends discoveries in plant physiology and biology with phenomenology and ecocritique to create what philosopher Quentin Hiernaux has described as a "new philosophy of vegetal life," one that rejects persistent notions of plant inferiority and dependence (Hiernaux 2019). In his 2016 book *Plant Theory: Biopower and Vegetable Life*, Jeffrey T. Nealon suggests that "Plants [...] are quickly becoming the new animals" (xiv). Since then, the field has only continued to expand its reach, knitting together philosophy (Marder 2014, Coccia 2016), cultural theory (Myers 2019, 2017a, 2017b, Sandilands 2019, 2016) and literary studies (Meeker and Szabari 2020), to name only a few, in an investigation of the plant as a subject that both withdraws from human modes of perception and engages them.

While this move towards the vegetal world is often positioned as a new development, the entanglement of humans and plants is an ancient one. In Western contexts, the vegetal body has stood as a figure for human modes of embodiment for centuries, if not millennia, even as the nature of this figuration has changed significantly over time. The ecologies that bind plants and humans together are as much cultural as they are biological. They include exchanges between texts and images as much as they do interactions among animate beings and organisms, and they have a history that predates that of the modern scientific study of plant and human encounters. But the enmeshment of humans with plants (and vice versa) takes on new contours under modernity, specifically in the role that the plant comes to play in the elaboration of biopolitical theories of life, starting around the eighteenth century.[2]

While Western modernity is often described as having brought a *dis*connection from the organic world, the plant is an important figure for gender under biopolitics. Vegetal life has a critical role to play in speculative feminist fictions that attempt to reorganize the relationship between women's

DOI: 10.4324/9781003023319-4

bodies and the state, fictions that respond to and revise longstanding efforts to conceive of the cultivation of humans in plant-like terms. At a moment when feminist scholars are once again considering plants in order to envision a more just and joyous future, it is crucial to attend to the way in which the plant has *already* played a key role in shaping norms around gendered and raced human bodies, including in feminist speculations that predate the current plant turn. These norms retain connections to eugenics and white supremacism that do not evaporate when they are deployed for feminist ends. In short, the history of the human entanglement with the plant body has shored up white supremacism as much as it has opened new utopian vistas – and often both at once.

In the eighteenth century, the human-plant analogy gains currency as a way of representing how humans need to be manipulated and formed in order to participate in society. In Jean-Jacques Rousseau's famous opening to the *Émile*, the young child is compared to a young shrub. "Plants are shaped by cultivation, and men by education," Rousseau writes in the opening paragraphs of the treatise (Rousseau [1762] 1979, 39). And, a bit later,

> The plant, set free, keeps the inclination it was forced to take. But the sap has not as a result changed its original direction; and if the plant continues to grow, its new growth resumes the vertical direction. The case is the same for men's inclinations. (39)

While the origins of this trope are classical,[3] the tendency to liken human children to plants, amenable to pruning, grafting and training, intensifies and spreads with modernity. Influenced by Rousseau among others, German educator Friedrich Froebel founded the first *kindergarten* –a garden of children – in 1837. Froebel had a strong interest in forestry and herbalism and included gardening in his curriculum.[4] The "naturalness" of the comparison of humans to plants is here so embedded in the project of education as to be almost invisible now, or at least unremarkable. In the *kindergarten*, acculturation of children comes to seem inevitably and organically linked to the cultivation of vegetal life.

How does gender enter into this ecology, where humans connect with plants physically and metaphorically? (Both Froebel and Rousseau believe in the salutary effects, for humans, of interactions with actual plants.) The comparison of feminine bodies to vegetal ones has often stressed the ornamental passivity of both women and plants. But in a context where the vegetality of the human subject is being remobilized as a mode of responsiveness to discipline, the seemingly feminine dimensions of plant life take on a more active cast. In the eighteenth century, plants also begin to play a role in efforts to envision technologies for new, more efficient, and more pleasurable modes of human reproduction, as in Vincent Miller's[5] brief satirical text entitled *The Man-Plant, or Scheme for Improving and Increasing the British Breed* ([1751] 1752), which appears as a ludic response to Julien Offray de La

Mettrie's (equally ludic) *Man a Plant* ([1748] 1994). Miller imagines transplanting human babies into artificial uteri and growing them in a hothouse, thereby lessening the travails of women, heightening their pleasure in sex, and enhancing the fertility of the British race. Viewed from the perspective of the twenty-first century, Miller's vision seems to anticipate Zygmunt Bauman's argument that the "gardening culture" of modernity involves from the outset a pruning and culling of human subjects for horrific genocidal ends, culminating in the Shoah (Bauman 1989). Bauman reads the situation of plants in the garden – in which vegetal bodies are subject to meticulous planning and manipulation – as a powerful metaphor for the control of human subjects that the modern state demands. Bauman's critique of bureaucratic rationalization, for which the garden becomes the image, renders the figure of the plant an objectifying and dehumanizing one. Miller's text, on the other hand, stresses the double-edged nature of this same trope, which strips humans of their privileged status among other life forms at the same time as it lends them a new energy: that of an amazingly prolific vegetal life. Miller's image of babies grown in hothouses mobilizes the powerful reproductive capacities of the plant, long understood as particularly apt at generation. Instead of operating only as figures for totalitarian violence, Miller's plant-wombs also tap into an extraordinary and inhuman force that may be put to work to expand and enrich human happiness generally and women's pleasure specifically. In Miller's hothouses as in Froebel's *kindergarten*, vegetalized human bodies are *both* exploited *and* productively deployed. Plant-humans represent at once the limit of human sociability and the generative transformation of human social life.

The image of the human as plant thus serves not just to objectify but to animate groups of people in new ways. Plants are exquisitely vulnerable and astonishingly powerful at one and the same time: they simultaneously withdraw from and instantiate human schemes for cultivation and acculturation. It is this double function of the modern plant – as victimized, often femininized body *and* source of vital and revitalizing energy – that is taken up by the tradition of feminist thought that this chapter will investigate in its raced and gendered implications. Feminist speculative fiction in the United States undertakes a reworking of the biopolitical plant – with its longstanding ties to femininity – as offering a possible exit from patriarchal oppression. In this context plant power is often explicitly made to operate in the service of white women's social and political agency. The doubleness of the plant reappears here – allowing an author like feminist Charlotte Perkins Gilman (1860–1935) to articulate a regime of white supremacy, in which white bodies are more "trainable" than others, with an argument on behalf of motherhood as the apogee of acculturation. In attending to vegetal being today, the entanglement of the plant – including the feminist plant – with this racist biopolitical legacy cannot be forgotten. Plant power is neither innocent nor neutral, and its history tracks and intersects with that of modernity more broadly.

Gilman and the Feminist Plant

Some decades before Gilles Deleuze and Félix Guattari call for readers to "follow the plants" (1987, 11) in their rhizomatic intertwinings, US author Charlotte Perkins Gilman revisits the plant-human analogy as an object of speculative revision and recuperation for modern feminism. Gilman remains best known for her short story "The Yellow Wall-Paper," first published in 1892 and reissued by Elaine R. Hedges and the Feminist Press in 1973. A key text for Second Wave US feminism and by-now a canonical entry in anthologies of US literature, "The Yellow Wall-Paper" depicts the gradual psychological unravelling of its protagonist following the birth of her child and her confinement to an upstairs room during a summer "rest cure". An influential social reformer and lecturer during her day, Gilman authored feminist theory and philosophy in addition to fiction and poetry, an expansive body of work that includes, most importantly for the purposes of this essay, a loosely connected trio of utopian novels of which the second, *Herland* (1915), has received the most critical attention. Although her work fell mostly out of print shortly after her death, Gilman's legacy was subject to a revival in the second part of the twentieth century thanks in large part to the efforts of feminist critics writing in the 1970s and 1980s, including Ellen Moers, Elaine Showalter, and Sandra Gilbert and Susan Gubar.[6] More recently, Gilman's corpus has been discussed in the context of ecocritique and environmental studies, in particular by Paul Formisano (2016), whose article on *Herland* places Gilman's utopian fiction in the context of California's Garden Movement and Southern California garden culture, and by Amanda Graham (1998), Janna Knittel (2006), and Susan Stratton (2001), who have examined Gilman's work as part of a larger tradition of ecofeminism.

Over time, scholars and historians have increasingly acknowledged the extent to which Gilman's interrogations of patriarchal culture and politics in the United States remain deeply and inextricably entangled with white supremacist ideology.[7] While Gilman gives a devastating analysis of the negative effects on white women of patriarchal society, eugenicist theories of evolutionary biology have a strong and persistent influence on her feminism that cannot be subtracted from her critical approach but instead constitute one of its enabling conditions. As Kyla Schuller (2018) writes in her analysis of "impressability" and white womanhood in nineteenth-century feminism:

> The claim that race is an 'aspect' of evolutionary feminisms, one of its discrete components that rigorous historical methods would best relegate to the side [...], manages to miss the significance of its own topic. The evolutionary notion of the distinct sexes of male and female, understood as specialized divergences in physiology, anatomy, and mental function that only the most civilized had achieved, was itself a racial hierarchy.
> (Schuller 2018, 105)

As Schuller's analysis explores, theories of femininity (and feminism) are inseparable from racist hierarchies in Gilman's work. Schuller's study affirms what Dana Seitler, in her article "Unnatural Selection: Mothers, Eugenic Feminism, and Charlotte Perkins Gilman's Regeneration Narratives," calls the "mutually constitutive" relationship between feminism and eugenics during this period, "each inextricably rooted in the constitution of the other" (2003, 64). While Gilman's racist and xenophobic tendencies are well-documented in both her public and private writings,[8] especially in her 1908 essay, "A Suggestion on the Negro Problem," and her 1923 piece, "Is America Too Hospitable?" these characteristics appear in Gilman's utopias as well, often in the form of an emphasis on pedigree and genetic purity.

Equally key to the elaboration of Gilman's utopian biopolitics is the figure of the plant, through which the dismantling of bourgeois patriarchy encounters the possibility of a withdrawal from the hybridity and racial mixing produced by conventional modes of acculturation and human reproduction. In *Herland*, Gilman makes clear the importance of vegetal modes of life for her reconfiguration of the role of women in racist regimes of reproduction. In this sense she is the legatee of Miller, who sees plants as a previously untapped source of reproductive power that can be channelled into projects of national and racial "improvement". For better and worse, then, Gilman can be said to inaugurate a plant turn of sorts within feminist science fiction and fantasy, but one that takes up where Miller's satirical text leaves off, rather than radically diverging from this inheritance.

Plant power is thus at the heart of Gilman's utopian fictions, but it ultimately functions in the service of white supremacy rather than in opposition to it. In *Herland*, Gilman harnesses the flexibility, adaptability and reproductive energies of the plant – its capacity to operate as an efficient technology of biopower – for the preservation of white women's supposed racial superiority. While elsewhere in her work, including most notably in the short story "The Giant Wistaria" (1891), the plant serves primarily to destabilize bourgeois patriarchal institutions, in *Herland*, and to a somewhat lesser extent in *Moving the Mountain* (1911) and *With Her in Ourland* (1916), Gilman's two other utopian novels from the same period, the specific capacities and drives of vegetal life animate a racially homogenous society of women. Gilman's corpus is thus a testament to the ability of the plant as figure to oscillate between destabilizing and consolidating social norms and to function in human contexts as both an agent and an object. Plant life exceeds and unsettles the patriarchal structures that bind white women within bourgeois family relations at the same time as it energizes white female bodies within a new regime of maternal reproduction. In this way the plant manages to adapt to human desires while retaining a force of its own that resists full incorporation into human institutions. In her Gothic narratives, including not just "The Giant Wistaria" but "The Yellow Wall-Paper," Gilman builds on speculative fantasies of plant life as disengaging from human patriarchal norms while still retaining an uncanny connection to human emotions. In these early stories, the plant works first of all to upend conventional assumptions about reproductive life and kinship bonds. In her utopias, on the other hand, the plant explicitly becomes a technology for racial "improvement". In its reproductivity and its responsiveness to its environment, the plant shores up racist modes of domination at the same time as it guarantees white women's empowerment; it comes to serve as a kind of master trope for both white supremacy and feminine independence, an old biopolitical entanglement already present in Miller's satirical text.

In Gilman's *Moving the Mountain*, the story of a man who returns to the United States of the 1940s after a 30-year period of isolation in Tibet, US culture and society have been transformed. The protagonist John Robertson marvels at the wealth, economy, efficiency and harmony of a society that has eliminated prostitution, poverty, unemployment and inequality. This is a society that is preoccupied with the "proper" cultivation of and respect for nature – vegetarianism is de rigueur, and hunting is frowned upon – and that tackles environmental problems, including pollution, deforestation and soil exhaustion, with the judicious application of scientific knowledge. In this reformed version of the United States, children's education is entrusted to "The Garden". As one character explains to John Robertson, "Yes, it's a regular Child Garden, where they are cultivated and grow! And they do so love to grow!" (Gilman 1911, 154). Here careful attention to the environment is coupled with a nascent sense of the shared interests and purposes linking human and vegetal life. The "garden of children" is, on the one hand, a commonplace trope that recalls not only Froebel's *kindergarten* but images of babies emerging from among rows of cabbages, as in Alice Guy-Blaché's pioneering short, *The Cabbage Fairy* ([1896] 1900), sometimes described as the world's oldest narrative film, and illustrations in seed catalogues from the same period. On the other hand, Gilman's first utopian novel gives the sense that this careful cultivation of children is of a piece with the careful management of the land: the human-plant analogy animates her utopia both in the practices that the latter enables and in the ontological assumptions that undergird it. Plants act as emblems of human growth and development at the same time as they remain objects to be managed and maintained.

Coming into focus here is not so much a separation between human culture and agricultural cultivation as the merging of these two realms into a zone in which the fruitfulness and energy of vegetal life shape the progress of human society and the bodies that inhabit it. Indeed, Gilman's vision echoes much older utopian projects, recalling Thomas More's *Utopia* (1516) in which both gardens and agricultural labour feature prominently. But Gilman is not just heightening the reproductive potential inherent in the plant but mobilizing this vegetal energy for the critique of bourgeois patriarchy from a white woman's point of view, as an eventual impetus for racial "progress". The plant becomes the image of the human being/woman who can adapt flexibly to circumstances – demonstrating the "impressability" that Schuller (2018) argues is a key component of white bourgeois womanhood in the nineteenth-century United States[9] – while still keeping her distance – effectively withdrawing – from patriarchal culture as a uniquely human mixture of bodies, affects, practices and beliefs.

Instead of marshalling the plant for a political project of hybridity or difference-in-vulnerability, as contemporary plant theorists often seek to do, Gilman enlivens a feminine vegetality as part of the cultivation of a "garden" of white womanhood. This vision comes to fruition in *Herland*, but there are hints of it already in "The Sweet Uses of Adversity," which appeared in Gilman's first book of poetry, *In This Our World*, published in 1893. This poem describes the "Norway birch" (*Betula pubescens*) in all its beauty and majesty, growing around the fjords in the summer: "[f]ull fed in all its nature needs/and showing/That nature in perfection by its growing" (1893, 95). But while the birch seems to appear in all its "perfection" by the Norwegian fjords, it is capable of change and adaptability. If the reader were to "follow the persistent tree/To the limit of endless snow – There you may see what a birch can be!" (96). In this more hostile environment, the tree faces difficult conditions, but nonetheless manages to survive and to preserve its specific mode of life. As Gilman writes, "Strong is the tree and rough, /It lives, and that's enough" (ibid.). As it turns out, the aesthetically appealing birch of summer misses, in its very beauty, the challenges of adaptation – of confronting and surmounting an environment that is not obviously suited for it. The poem concludes: "That silver monarch of the summer wood, /Tall, straight, and lovely, rich in all things good; Knew not in his perversity/The sweeter uses of adversity!" (ibid.). Throughout the poem, Gilman elaborates on the dual "nature" inherent in the plant. On the one hand the birch is a figure of integrity and harmony with the environment, perfectly at home in the place that it inhabits and bodying forth the more profound perfection of nature itself. On the other hand this tree displays the ability, in the second part of the poem, to "self-cultivate" and to adapt to an environment that does not welcome it. In this harsh and unforgiving context it serves not so much as an example of aesthetic perfection as of "life" itself. Pleasure is available in both modes, but its source has changed. The force that renders the plant capable of triumphing over the vicissitudes that assail it is itself a mode of "sweetness," but one that suggests the power to work against or with its circumstances rather than to remain captive to them. The plant in the final image from the poem is not passive but active – possessed of its own ability to encounter and to transform itself in relationship to the environment that surrounds it and gives it shape. In a sense it "self-cultivates," collapsing into itself two opposed notions of vegetal growth: the first, as a purposeless and unconscious force that is continuous with the atmosphere in which the plant lives, and the second, as a conscious "training up" of the vegetal body in order to suit a particular objective (usually a human one, albeit not in the case of this poem). In its duality as both beautiful, shapely poetic object and active subject, the birch tree has an uncanny kinship with the women of *Herland*, who not only live in isolation but reproduce by means of parthenogenesis, a mode of generation common both to plants and to what is described by the character Jeff ("born to be a poet, a botanist – or both" [Gilman 1999, 4]) in the novel as "'some rather high forms of insect life'" (53).

Some critics have seen in *Herland* a technocratic vision of the ideal management of nature by humans. Andrew G. Christensen, in his reading of *Herland* within the tradition of the scientific utopia (again stretching back at least to Thomas More), finds in Gilman's work a depiction of environmental engineering so thorough-going as to be characterizable as a form of "ecocide". In Christensen's reading, "Culture is no longer opposed to nature because nature has been thoroughly brought under the control of culture" (2017, 291). He thus resists the implication that Gilman anticipates a kind of ecofeminist environmentalism in which humans and nature exist on a harmonious continuum, preferring instead to position Gilman's writing within a longer history of scientific management of nature for specifically human utopian ends. Christensen is not wrong to stress Herland's connection to an early modern utopian garden that represents not the embeddedness of humans in nature but the exploitation of the natural world by humanity. What his genealogy leaves out, however, is the gradual development of a vegetal biopolitics in which the plant itself is an active technological force within a modern world that operates to render humans more plant-like by positioning and "cultivating" them as vegetal bodies in their right. It is this legacy, one that intersects with the scientific utopian tradition that interests Christensen but is not fully identical with it, that Gilman explicitly redeploys. The denizens of Herland are a feminist instantiation of woman-as-plant: self-cultivating along the lines of the Norwegian birch and flamboyantly, triumphantly reproductive in their vegetal nature.

In a fascinating article that takes up the question of Gilman's biopolitics and stresses the relationship between Herland as utopian garden and Bauman's gardening culture, Agnes Malinowska (2019) suggests that Herland is built around the simultaneous acknowledgement and disavowal of what she describes as a "fungal femininity" (268) – a kind of helpless, ceaseless reproductivity that evokes the multiplication of microorganisms. For Malinowska,

> The utopian nations that Gilman envisions in *Herland* and *Moving the Mountain* (1911) are themselves 'fungal' in their sweeping collectivity, but these social organisms are fully *rationalized* ones. The future women that govern them subdue nature's unruly power through a meticulous application of efficiency principles.
>
> (Malinowska 2019, 268)

I would suggest that this "rationalization" of women's reproductive function – its channelling and management into a collectivity that is able to meet all of its needs with nature, as in "The Sweet Uses of Austerity" – is not so much the repression of women's animal being as it is an expression of their vegetal one. If, as Malinowska points out, "the Herlanders have almost completely purged their country of its nonhuman animals," they have at the same time lovingly cultivated the plants that sustain them (275).

Herland is narrated from the point of view of Vandyck Jennings, a young man with an interest in sociology who accompanies his two friends, Jeff and Terry, on a scientific expedition "up among the thousand tributaries and enormous hinterland of a great river" (Gilman 1999, 6). The narrator Van describes the "Woman Land" that the three men discover on their journey as "a land in a state of perfect cultivation, where even the forests looked as if they were cared for; a land that looked like an enormous park, only it was even more evidently an enormous garden" (1999, 24). While the women of Herland beautifully tend and manage their environment – the eventual love interest of Van and protagonist of *With Her in Ourland*, Ellador, is a forester – they do so *as if they were plants themselves.* Their love for one another and the world that surrounds them, their parthenogenetic mode of reproduction, their collectivism, even their withdrawal from the "outside," all suggest a mode of life that is vegetal in both its active and its passive dimensions. Like the summer birch in the first stanzas of "The Sweet Uses of Austerity," they exist as emblems of natural beauty and integrity, "[f]ull fed in all [their] nature needs" (1893, 95). Similar to the stunted but "tough" little trees at the poem's end, they meet the challenges of their environment by adapting and self-cultivating, always motivated in their actions by love for one another.

As the novel proceeds, the reader discovers that the women of Herland were originally members of a society that included both men and women. Thanks to a series of political and environmental catastrophes, all of the men who initially lived among them were killed; yet at this point, rather than dying out, the surviving women flourished through the miracle of parthenogenesis, a mode of reproduction that Gilman describes as subject to conscious control by the women themselves. The Herlanders thus have come to worship mothers and maternity as the source of life itself. Ellador describes to Van

> [t]heir great Mother Spirit [who] was to them what their own motherhood was – only magnified beyond human limits. That mean that they felt beneath and behind them an upholding, unfailing, serviceable love – perhaps it was really the accumulated mother-love of the race they felt – but it was a Power.
>
> (Gilman 1999, 121)

In this glorification of a maternal ideal, Herland is perhaps less reminiscent of the evolutionary dynamics of animal life forms (whether human or not) and more profoundly evocative of what evolutionary biologist Henry Drummond, whose *The Ascent of Man* (1894) serves as one of Gilman's intertexts, calls "The Struggle for the Life of Others" (as opposed to "the Struggle for

Life" *tout court*), one of whose principle avatars is the vegetal seed. In *The Ascent of Man*, Drummond writes:

> Every plant in the world lives for Others. It sets aside something, something costly, cared for, the highest expression of its nature. The Seed is the tithe of Love, the tithe which Nature renders to Man. When Man lives upon Seeds he lives upon Love. Literally, scientifically, Love is Life. If the Struggle for Life has made Man, braced and disciplined him, it is the Struggle for Love that sustains him.
>
> (Drummond 1894, 296)[10]

It is this sustaining vegetal commitment to love – rendered as a form of maternity through parthenogenesis – that is exemplified by the women of Herland, who are fully devoted to one another in their investment in motherhood as the principle that preserves their society.

Indeed, the Herlanders' very mode of reproduction is more vegetal than it is animal, the existence of parthenogenetic animals notwithstanding. When the three men begin to discuss the population and management of Herland with the women tutors assigned to them after their arrival, they express amazement not only at the absence of men but at the variations within the Herland population (which is nonetheless uniform in its beauty, strength, and intelligence). Terry suggests that parthenogenetic reproduction would render the whole population "as alike as so many ants or aphids" (1999, 84). For Terry, the existence of "visible differences" among the women suggests that sexual difference must also exist in their society – that "there must be men – somewhere" (ibid.). The women respond to Terry that their model in this regard is not the world of insects but that of plants, which allows for the possibility of mutability (and training) in the absence of sexual reproduction.

> But when we asked them, in our later, more intimate conversations, how they accounted for so much divergence without cross-fertilization, they attributed it partly to the careful education, which followed each slight tendency to differ, and partly to the law of mutation. This they had found in their work with plants, and fully proven in their own case.
>
> (Terry 1999, 85)

The text goes on to specify that, despite their differences, the women are more similar to one another than is the case for the populations with which the men are most familiar, in that the Herlanders are entirely without disease or "excessive types". As a race, in Gilman's portrayal, they are the picture of health and beauty, so that their individual differences have no effect on the integrity of the collective body. It should come as no surprise that the children of Herland – all girls – are a little later on described in vegetal terms. As Van puts it, "Those nation-loved children of theirs compared with the average in our country as the most perfectly cultivated, richly developed roses compare with – tumbleweeds. Yet they did not *seem* 'cultivated' at all – it had become a natural condition" (1999, 78). This is the vegetal model of education, embodied by the *kindergarten*, as not only a figural but a literal garden of humans. The children of Herland take the vegetal talent for seemingly invisible and effortless self-cultivation and transform it into a national ideal. There is undeniably environmental management occurring in Herland, but rather than a separation between humans and the natural world that they are called upon to regulate and exploit, these Herland practices suggest the extraordinary power of plants to cultivate themselves – to mutate, change and adapt to their circumstances while remaining true to the nature that inhabits them.

It is this latter nature that Gilman very clearly deploys on behalf of a white supremacist vision. The Herland women are not only racially distinct from the tribal people who live closest to them (without for all that being able to access Herland, which is cut off from the rest of the land that surrounds it) but from the people of the United States, who are described as indiscriminately and helplessly reproductive (here along the lines of the "fungal femininity" invoked by Malinowska [2019]). The Herlanders have withdrawn from the vicissitudes of modern human culture while still being able to adapt to their environment, both through the use of cultivated mutation and through the education and training of their bodies and minds. Neither of these processes is incompatible with a vital, animate and biopolitical vegetal life. In fact, as I have described in the introduction, they have been linked to the exemplary operations of this life at least since the eighteenth century. Gilman marshals this "enlightened" modern biopolitics for her feminist project, all the while reinforcing its embeddedness in a deeply racist sense of collective destiny and improvement. Viewed through this prism, Herland becomes a society of plants in its most superficially appealing as well as its most viscerally disturbing aspects.

The Contemporary Woman-as-Plant

Gilman's insight into the utopian promise of the active plant is reworked and reconsidered throughout the twentieth century and into the twenty-first century by authors of feminist and "feminine" speculative fiction – to use Belgian author Anne Richter's turn of phrase (Richter 1995) – who return to the trope of woman-as-plant from many different vantage points.[11] Richter's short story "The Sleep of Plants," from her collection *Les Locataires* (1967), describes a young woman who literally "pots" herself (like a houseplant) in order to passively resist the influence of her mother and to avoid marriage to her fiancé. Later on, in her novel *The Vegetarian* (2007) and the short story "The Fruit of My Woman" (1997), South Korean author Han Kang will re-animate the trope of the woman-as-plant to explore feminine vegetality as a response to the traumas of late capitalism. In Iranian author Shahrnush

Parsipur's novel *Women Without Men* (1989), a woman becomes a tree after planting herself in her family's garden. Perhaps the most powerful late-modern Anglophone reconfiguration of the woman-as-plant trope within speculative fiction occurs in Octavia Butler's Earthseed series, which might be understood as a direct response to and repudiation of the racist biopolitics of an author like Gilman, all the while retaining the seed metaphor as its central animating figure.

In *Parable of the Sower*, the first book of the Earthseed saga, Butler depicts protagonist's Lauren Oya Olamina's struggle for survival as a Black woman in a near-future world deeply hostile to human life. The daughter of a Baptist minister, Lauren is forced to flee her pillaged home into the dystopian landscape of post-apocalyptic Los Angeles. Collecting around her a small band of fellow travellers, Lauren at once fights for the survival of her group and establishes a new religion: Earthseed, based on the doctrine that "The only lasting truth is Change" (Butler 1993, 13). As in Gilman's speculative utopias, plant life functions here as a model for understanding the potential inherent in human life, but Butler restructures this connection in order to argue on behalf of rather than against the possibility of interracial collaboration, care and love. Most importantly, the trope of the seed as both vulnerable *and* powerful allows Lauren (and by extension, the reader) to bear witness to the history of violence that saturates Black experience in the United States. At the same time, the vitality of the seed becomes a testament to the strength and beauty of Black survival. In Butler's novel, the walls of the utopian garden have fallen and been torn down; there is no shelter from violence available to Lauren and she cannot withdraw from the effects of chaos. It is the vegetal seed that serves as a primary model of resilience and ongoing survival in a deeply hostile world. As a trope, it denotes both movement and transformation. Lauren reflects:

> Well, today, I found the name, found it while I was weeding the back garden and thinking about the way plants seed themselves, windborne, animalborne, waterborne, far from their parent plants. They have no ability at all to travel great distances under their own power, and yet, they do travel. Even they don't have to just sit in one place and wait to be wiped out [...] I am Earthseed. Anyone can be. Someday, I think there will be a lot of us. And I think we'll have to seed ourselves farther and farther from this dying place.
>
> (Butler 1993, 87–88)

Lauren's Earthseed may not move of its own volition, but it moves nonetheless. And in moving it reshapes a community of hardy, collaborative, suffering but joy-seeking humans whose existence both testifies to the history of racist violence in the US and suggests a possible exit from it.

The ambivalent legacy left by the corpus of feminist plant fiction that includes both Gilman and Butler provides an important context for considering the ambitions and limitations of today's plant theory, which comes back to the plant as a potential interlocutor for those humans in search of a more just world. Theorists of critical plant studies present plants as both embedded in human modes of signification and capable of bypassing or eliding all-too-human forms of violence and harm, in ways that humans might learn from or respect rather than seek to destroy. Moreover, plant theorists are once again turning to plants as ambiguous and sometimes ambivalent figures of femininity and feminism, as in Luce Irigaray's collaborations with Michael Marder in *Through Vegetal Being: Two Philosophical Perspectives* (2016) and in the recent article by Prudence Gibson and Monica Gagliano on "The Feminist Plant" (2017), which focusses on the water lily. For Marder and Irigaray as for Gibson and Gagliano, plant life inhabits and responds to a modernity that seeks to degrade and denature it; understood in the specificity of their mode of life, plants can offer a vulnerable yet powerful antidote to the late capitalist economies that work to subjugate them. As Gibson and Gagliano put it

> As with most plants, the water lily has been marginalized and treated as unequal or ontologically less relevant in the past by humans. Vilified for being too passive, too silent, disparaged for having inferior abilities and condemned for being immobile, the water lily nevertheless defies these anthropocentric denigrations. At this point of crossover between nature and culture, and as an emerging interpretation of that relation, the water lily is a feminist.
>
> (Gagliano 2017, 126)

Gibson and Gagliano stress the ways in which water lilies "reflect a feminist, even a transgender or gender-fluid, life due to their adaptive option to self-reproduce" (130). They see this mode of life as countering or working against unjust human social norms, including the objectification of bodies under capitalism and the fetishization of white cisgender heterosexuality. In this context, they seek to highlight the autonomy of the water lily as a mode of being, valuable in and for itself, that might provide an example for humans who desire a feminist remedy to the trauma of the modern world.

The struggle for a more ecologically attentive relation to plants is both necessary and important, but feminists cannot and should not forget the extent to which vegetal life has a long history of shaping gendered norms *within* modernity rather than outside of it. Indeed, as I have tried to show here, plants have often been deployed and hailed as an exemplary form of biopolitical life by feminists and non-feminists alike. Moreover, the desire to activate plant power as a progressive means of improving human existence – the emanation of a uniquely malleable and autonomous life form, especially in its nonhuman modes of reproduction – has worked in tandem with a racist biopolitics that at best seeks with the withdrawal from human culture to escape a reckoning with the specific traumas of human history. At worst this deployment of the plant to critique patriarchal regimes of reproduction shores up a racist vision of genetic purity. Recognizing the agency and vitality of other life forms, and vegetal life forms in particular, is not the same as rendering justice. The desire to recuperate a pure vegetality from the ravages of modern culture – to somehow be in communication with plants from outside of or beyond the wrecked world that we inhabit with them – can be troublingly proximate to the desire to elide or slip past the violence done in the name of human being's differences from one another. It can even, as it does in Gilman's case, conspire with the regimes that depend on this violence. While it is tempting to come back to the plant as both a figure of femininity that escapes the traumatic entanglements of modernity (which include of course intra-human entanglements in relationships of harm *and* care), this understanding of the plant has its own human past. We still need to reckon with the work that the plant does *for* modernity before we can firmly situate it on modernity's outside or beyond. In this project, feminist speculative fiction can reveal the perils and the promise of the plant as a model for a better human life.

Notes

1 This term is coined by Michael Marder; his series of the same name was founded with Brill and Rodopi in 2013.

2 For a nuanced discussion of vegetal life in its relationship to contemporary biopolitical theory and practice, see Catriona Sandilands (2016). Jeffrey Nealon also highlights the biopolitical potential of plant life in *Plant Theory: Biopower and Vegetable Life* (2016).

3 See the first chapter of Marder (2014) for a discussion of the human-plant analogy in Plato (in addition to other classical philosophers).

4 For a discussion of Friedrich's attachment to botany and the use of gardens in his schools, including his first kindergarten in Bad Blankenburg, see Murcia and Ruiz-Funes (2020). Friedrich also develops at length an analogy between a human child and a kernel of corn in *Pedagogics of the Kindergarten: Ideas Concerning the Play and Playthings of the Child* (1861).

5 This name appears to be a pseudonym. The identity of the actual author has not been established. Thank you to Daisy Reid for bringing this text to my attention.

6 For more on the revival of scholarship on Gilman, see Dock et al. (1996).

7 On Gilman's "racial nationalism," see Weinbaum (2004).

8 On Gilman's racial prejudices, as they are expressed in both her published work and her private correspondence, see Knight (2000).

9 Schuller presents impressability as a prominent concept within American evolutionary science, moral philosophy and the uplift politics of the nineteenth century; in these contexts, the term refers to the capacity of an organism to actively respond to its environment.

10 Dana Seitler (2003, 77) notes that Gilman was also inspired by the work of Lester Frank Ward, a US palaeobotanist who became a sociologist.

11 On the trope of women-as-plant see also Swarnalatha Rangarajan (in this volume).

References

Bauman, Zygmunt. 1989. *Modernity and the Holocaust.* Ithaca, NY and New York: Cornell University Press.

Butler, Octavia E. 1993. *Parable of the Sower.* New York: Seven Stories Press.

Christensen, Andrew G. 2017. "Charlotte Perkins Gilman's *Herland* and the Tradition of the Scientific Utopia." *Utopian Studies* 28 (2): 286–304.

Coccia, Emanuele. 2016. *La vie des plantes: Une métaphysique du mélange.* Paris: Rivages.

de La Mettrie, Julien Offray. (1748) 1994. "Man a Plant." In *Man a Machine and Man a Plant.* Translated by Richard A. Watson and Maya Rybalka, 77–92. Indianapolis: Hackett Publishing Company.

Dock, Julie Bates, Daphne Ryan Allen, Jennifer Palais, and Kristen Tracy. 1996. "'But One Expects That': Charlotte Perkins Gilman's 'The Yellow Wallpaper' and the Shifting Light of Scholarship." *PMLA* 111 (1): 52–65.

Deleuze, Gilles and Félix Guattari. 1987. *A Thousand Plateaus: Capitalism and Schizophrenia.* Translated by Brian Massumi. Minneapolis: The University of Minnesota Press.

Drummond, Henry. 1894. *The Lowell Lectures on the Ascent of Man.* London: Hodder and Stoughton.

Formisano, Paul. 2016. "'It Had All Become a Natural Condition': California's Garden Movement, Land Eugenics, and Naturalization in Charlotte Perkins Gilman's *Herland.*" *Western American Literature* 51 (1): 71–100.

Froebel, Friedrich. (1861) 1897. *Pedagogics of the Kindergarten or, His Ideas Concerning the Play and Playthings of the Child.* Translated by Josephine Jarvis. London: Edward Arnold.

Gibson, Prudence and Monica Gagliano. 2017. "The Feminist Plant: Changing Relations with the Water Lily." *Ethics & the Environment* 22 (2): 125–146.

Gilman, Charlotte Perkins. 1893. "The Sweet Uses of Adversity." In Papers, 1846–1961, Folder 269. *In This Our World*, 1893 edition, with Charlotte Perkins Stetson's corrections; and proof of title page of 1895 edition. http://pds.lib.harvard.edu/pds/view/13725856 seq.107.

Gilman, Charlotte Perkins. 1908. "A Suggestion on the Negro Problem." *American Journal of Sociology* 14 (1): 78–85.

Gilman, Charlotte Perkins. 1911. *Moving the Mountain.* New York: Charlton Company.

Gilman, Charlotte Perkins. 1923. "Is America Too Hospitable?" *The Forum*, October: 1983–1989.

Gilman, Charlotte Perkins. (1916) 1997. *With Her in Ourland. Sequel to Herland.* Edited by Mary Jo Deegan and Michael R. Hill, 59–193. Westport, CT: Greenwood Press.

Gilman, Charlotte Perkins. (1915) 1999. *Herland.* In *The Yellow Wall-Paper, Herland, and Selected Writings*, 1–156. New York: Penguin Books.

Gilman, Charlotte Perkins. (1892) 1999. "The Yellow Wall-Paper." In *The Yellow Wall-Paper, Herland, and Selected Writings*, 179–196. New York: Penguin Books.

Gilman, Charlotte Perkins. (1891) 1999. "The Giant Wisteria." In *The Yellow Wall-Paper, Herland, and Selected Writings*, 167–175. New York: Penguin Books.

Graham, Amanda. 1998. "*Herland*: Definitive Ecofeminist Fiction?" In *A Very Different Story: Studies on the Fiction of Charlotte Perkins Gilman*, edited by Val Gough and Jill Rudd, 115–128. Liverpool: Liverpool University Press.

Guy-Blaché, Alice, dir. 1896. *The Cabbage Fairy.* Paris: Gaumont, 1900. Film.

Hiernaux, Quentin. 2019. Interview with Sonya Faure for *Libération.* https://www.liberation.fr/debats/2019/04/17/le-corps-intelligent-des-vegetaux-ebranle-notre-philosophie_1721953/.

Irigaray, Luce and Michael Marder. 2016. *Through Vegetal Being: Two Philosophical Perspectives.* New York: Columbia University Press.

Kang, Han. (2007) 2015. *The Vegetarian.* Translated by Deborah Smith. New York: Penguin Random House.

Knight, Denise D. 2000. "Charlotte Perkins Gilman and the Shadow of Racism." *American Literary Realism* 32 (2): 159–169.

Knittel, Janna. 2006. "Environmental History and Charlotte Perkins Gilman." *International Review of Science Fiction* 35 (96): 49–67.

Malinowska, Agnes. 2019. "Charlotte Perkins Gilman's Fungal Female Animal: Evolution, Efficiency, and the Reproductive Body." *Modernism/modernity* 26 (2): 267–288.

Marder, Michael. 2014. *The Philosopher's Plant: An Intellectual Herbarium.* New York: Columbia University Press.

Meeker, Natania and Antónia Szabari. 2020. *Radical Botany: Plants and Speculative Fiction.* New York: Fordham University Press.

Miller, Vincent. (1751) 1752. *The Man-Plant or, Scheme for Increasing and Improving the British Breed.* London: M. Cooper, at the Globe in Paternoster Row.

More, Thomas. (1516) 1984. *Utopia.* Edited and translated by Paul Turner. New York: Penguin Classics.

Murcia, José Pedro Marín and María José Martínez Ruiz-Funes. 2020. "Froebel and the Teaching of Botany: The Garden in the Kindergarten Model School of Madrid." *Paedagogica Historica* 56 (1–2): 200–216.

Myers, Natasha. 2019. "From Edenic Apocalypse to Gardens against Eden: Plants and People in and after the Anthropocene." In *Infrastructure, Environment, and Life in the Anthropocene*, edited by Kregg Hetherington, 115–148. Durham, NC: Duke University Press.

Myers, Natasha. 2017a. "Becoming Sensor in Sentient Worlds: A More-than-natural History of a Black Oak Savannah." In *Between Matter and Method: Encounters in Anthropology and Art*, edited by Gretchen Bakke and Marina Peterson, 73–96. London: Bloomsbury Press.

Myers, Natasha. 2017b. "From the Anthropocene to the Planthroposcene: Designing Gardens for Plant/People Involution." *History and Anthropology* 28 (30): 297–301.

Nealon, Jeffrey T. 2016. *Plant Theory: Biopower and Vegetable Life.* Stanford, CA: Stanford University Press.

Parsipur, Shahrnush. (1989) 2011. *Women without Men: A Novel of Modern Iran.* Translated by Faridoun Farrokh. New York: The Feminist Press.

Richter, Anne. 1995. *Le fantastique féminin d'Ann Radcliffe à Patricia Highsmith.* Bruksela: Éditions Complexe.

Richter, Anne. (1967) 2015. "The Sleep of Plants." Translated by Edward Gauvin. In *Sisters of the Revolution: A Feminist Speculative Fiction Anthology*, edited by Ann and Jeff VanderMeer, 131–136. Oakland, CA: PM Press.

Rousseau, Jean-Jacques. (1762) 1979. *Émile or On Education.* Translated by Allan Bloom. New York: Basic Books.

Sandilands, Catriona. 2016. "Floral Sensations: Plant Biopolitics." In *The Oxford Handbook of Environmental Political Theory*, edited by Teena Gabrielson, Cheryl

Hall, John M. Meyer, and David Schlosberg. Vol. 1. Oxford: Oxford University Press. doi:10.1093/oxfordhb/9780199685271.013.33.

Sandilands, Catriona A. H., with Olga Cielemęcka and Marianna Szczygielska. 2019. "Thinking the Feminist Vegetal Turn in the Shadow of Douglas-firs: An Interview." *Catalyst: Feminism, Theory, Technoscience* 5 (2). https://catalystjournal.org/index.php/catalyst/article/view/32863.

Schuller, Kyla. 2018. *The Biopolitics of Feeling: Race, Sex, and Science in the Nineteenth Century*. Durham, NC: Duke University Press.

Seitler, Dana. 2003. "Unnatural Selection: Mothers, Eugenic Feminism, and Charlotte Perkins Gilman's Regeneration Narratives." *American Quarterly* 55 (1): 61–88.

Stratton, Susan. 2001. "Intersubjectivity and Difference in Feminist Ecotopias." *FEMSPEC* 3 (1): 33–43.

Weinbaum, Alys Eve. 2004. *Wayward Reproductions: Genealogies of Race and Nation in Transatlantic Modern Thought*. Durham, NC: Duke University Press.

3 The Arboreal Feminine

An Analysis of Affect and Activism in Two Ecofeminist Re-Enchantment Narratives from India

Swarnalatha Rangarajan

Decolonial feminist epistemologies from Global South environments involve a confluence of indigenous feminisms and ecofeminism, and embrace a diverse range of issues spanning from geopolitics to body territory (Kozhisseri 2020, 139). For Vandana Shiva, the re-enchantment of nature is synonymous with the recovery of *prakriti*, a term in Hindu cosmology for "the living force that supports life", which is inextricably bound up with "person and nature (Purusha-Prakriti) [...] a duality in unity" (1989, 140). This recovery, according to Shiva, will offer "an intellectual and political challenge" to patriarchally driven projects of domination and uneven development that endanger both nature and women (14).[1] This essay argues that the environmental justice discourse of Global South ecofeminism also represents a call for re-enchantment, a process characterized by reciprocity and systems thinking. Re-enchantment in this context involves a paradigm shift regarding the "reconceptualization of difference", which is achieved by establishing a "complex web of new connections" between the human and other-than-human that dissolves oppressive structures of human domination over the biophysical world and other related forms of subjugation of women, children, people with disabilities and subaltern groups (Kurth-Schai 2005, 153).

Peter Linebaugh traces the etymology of enchantment to the French "chanter" meaning "to sing", and defines a song as "the call for enchanting the world, for singing creation into being, (that) is both rhapsodic and prophetic" (Linebaugh 2019, xvii). The new world forged through this creative singing is linked to community life and the struggle that invariably accompanies the act of creation. Linebaugh describes the work that goes into the process of re-enchantment as a revolutionary project that is inseparable from the commons and the refusal of privilege: "Read, study, think, listen, talk, and with others act, by which I mean fight [...] the new world is around us, it is about us, and only our struggle can bring it into existence and re-enchant it" (xvii). In addition, James William Gibson describes the "culture of re-enchantment" as "a new covenant

DOI: 10.4324/9781003023319-5

between people, land, creatures" (2009, 12) which refuses to reduce nature to a set of objects infused with utilitarian purpose. The sacralization shift also embraces "dark ecology" (Morton 2016), in which the attentive, compassionate gaze is directed at contaminated and toxic landscapes with the objective of transforming them through sustained action. Although re-enchantment may not be a substitute for environmental justice struggles, it gives rise to an extended moral community that acts with an awareness of the "relational, total-field image" (Naess 1973, 95) in which subjects "are not conceived of as 'things in themselves'" but in terms of their relationships to larger gestalts, the understanding of which can serve as a corrective for Anthropocene neglect. This chapter, while acknowledging nature's innate capacity to enchant, defines re-enchantment as a conscious relational strategy that can help the reader/community to nurture affective bonds with non-human nature. In what follows, I analyse two ecofeminist narratives that make a strong case for the re-enchantment of nature in the face of growing ecological devastation. The first text is the life story of the tribal leader C. K. Janu, the leader of the Muthanga land struggle by the dispossessed Adivasis (a collective term for the many indigenous peoples of India) of Kerala. C. K. Janu's autobiography, titled *Mother Forest: The Unfinished Story of C. K. Janu* (2004), evocatively describes her childhood and life in the forest and her subsequent political awakening. This life story is an oral narrative infused with the worldviews of her community and the everyday rituals of life in the forest, which can be considered a kind of "world interpretation" where the interaction of humans with the non-human world contributes to stories that are vital to the well-being of the community. The environmental justice struggle that ensues with the loss of land and expulsion from this maternal "*oikos*" can be seen as a reaction against the disenchantment of modernity and a bid to recover what Deborah Bird Rose calls an "eco-place": "an ancestral footprint – a living sign of past action" (2004, 175) which is characterized by interactions between land and actors, both human and non-human. The second text analysed in this essay is *How I Became a Tree* (2017) by Sumana Roy, a mainstream writer. The work is a striking example of ecofeminist re-enchantment, since it reconceives nature as innately innovative, spiritual and poetic and reciprocally connects the human mind to nature's creativity. However, it is also informed by a strong vein of environmental ethics, which connects it to C. K. Janu's environmental justice narrative. Read in tandem, these texts offer a trans-scalar[2] perspective on the ecologies of gender in India.

Woman as Forest: The Re-Enchantment of the Commons in C. K. Janu's Life Narrative

C. K. Janu hails from the Wayanad district of Kerala, a south Indian state known for its natural beauty and high human development index. Kerala also has the highest tribal population in the country, of which about 30 per cent is landless (Bijoy and Raman 2003). C. K. Janu's parents were indentured Adivasi labourers belonging to the Adiya ("slave") tribe, one of the 35 tribe clans of Kerala who occupied the lowest tier of the agricultural hierarchy. Most of the tribals were imported to this region in the nineteenth century in order to help with paddy cultivation. They were a largely self-sufficient community, who followed traditional methods of cultivation such as "slash and burn" and grew traditional crops like millet, sorghum and paddy. However, their living conditions deteriorated when non-Adivasi farmers moved to Malabar from Central Travancore after the Second World War and encroached on sizeable tracts of tribal land used for shifting cultivation (Bhaskaran 2004, vi). The Kerala Land Reform Act of 1957, considered to be progressive, was not successful in solving the longstanding issue of the transfer of tribal lands since it favoured the non-tribals who took tribal lands on short-term leases and later claimed ownership.

After the 1980s, the agricultural sector in Kerala witnessed a steady decline of productivity due to factors like erosion and commodification of land. Added environmental stress caused by development projects, eco-tourism and social forestry reduced the Adivasis to a state of abject servitude. The tipping point of the Muthanga rebellion spearheaded by C. K. Janu came in the wake of the deaths of Adivasis from starvation in July 2001, when 300 Adivasi families put up makeshift refugee huts outside the chief minister's office in Thiruvananthapuram and demanded their right to sources of livelihood. Commenting on the qualities that set the Muthanga rebellion apart from mainstream protests, Ravi Raman observes, "In beauteous contrast to the undemocratic and uncultured slogans shouted by the usual political agitators, the Adivasis swayed to the chanting of songs of anger, peace, and love" (2002, 917). *Mother Forest* is a counterculture narrative, which appeared originally in Malayalam as *Janu: C. K. Januvinte Jeevitha Katha*. It comprises Janu's reminiscences, as told to and written by Bhaskaran, of the history of the Adivasis and their identity and culture, which is closely interwoven with the life of the forest. Although the English translation does not recreate the richness of the oral tradition, the translator takes care to foreground Janu's relational worldview and her deep relationship with the forest, which drives her political activism. As he observes,

> she calls the struggle for land to Adivasis (as tribals are called in Malayalam) not a political one, but a struggle against another culture (meaning the normal Kerala culture). She says retrieval of land will help them retrieve their culture. Adivasis are not interested in political power. They just wish to be left alone to create their own methods of survival and development. (Bhaskaran 2004, viii)

The forest is not a backdrop in this life narrative but a living presence that provides the space for the marginalized to forge an autonomous self-identity.

Forests and their connotations of "wilderness" carry a negative valence in Western civilization. However, *aranya sanskriti* or the culture of the forest was once considered the bedrock of Indian civilization. The forest was seen as the great nurturer and embodiment of the feminine principle of nature, deified as the goddess Aryanyani, who bestowed both spiritual and material gifts. The environmentalist Nanditha Krishna observes that in Vedic times, "The forest was regarded as the highest form of cultural evolution. Human ability to merge with nature was the measure of cultural evolution" (2020, 169). According to Rabindranath Tagore, *aranya sanskriti* is the foundational "principle of Indian civilization" and "democratic pluralism" (Shiva 1989, 55). Vandana Shiva attributes the vibrancy of the forest culture to the everyday practices and beliefs of the tribal forest dwellers and points out that the colonial rule that erased indigenous systems of forest management also decimated *aranya sanskriti* in the process. The Indian Forest Acts passed in 1865 and 1878 gave the colonial government unlimited power over forest land, which resulted in the forests being declared state property and used as a resource base for timber. The Forest Act of 1927 further aggravated the situation by restricting the communal ownership of forests and adopting coercive methods to deal with tribal forest dwellers who, when faced with the loss of their subsistence economy, were forced to turn to other occupations. The postcolonial forest policy of 1952 exercised increased control over the tribals' private forest spaces, denied them the right of grazing and marginalized them further. The creation of Protected Area Networks enabled the construction of Adivasis as "encroachers" and facilitated the eviction of forest-dwelling tribes. However, the Forest Act of 2006 sought to redress the grave injustices meted out to the forest-dwelling communities since colonial times by taking cognizance of their rights to access forest produce, dwell on land and apply indigenous knowledge systems to conservation practices. Although there are controversies surrounding the implementation of this act, it has opened up a space where tribals can reclaim their rights and exercise local self-governance over the forest.

Janu's forest is a place of emergence that acknowledges and respects the myriad entanglements with the non-human as necessary for the survival of the human (Rangarajan 2016, 445). At the heart of the modernist vision of the world, in which everything has to be tamed, measured and understood from an anthropocentric perspective, lies what Weber refers to as "disenchantment" (Schroeder 1995, 228). Janu's resistant discourse refuses to instrumentalize the forest and its resources and articulates the language of symbiosis, which enables Janu to connect the maternal *oikos* of her childhood with the public sphere characterized by an emancipatory environmental agenda. Claiming a filial bond with the forest, Janu declares: "No one knows the forest like we do. The forest is mother to us. More than a mother, because she never abandons us" (Bhaskaran 2004, 5). The sensuous language offers rich snapshots of life in the forest that do not romanticize the place as one of primitive plenitude, devoid of the exigencies of everyday life, but rather understand it as an *oikic* space mediated by labour. Janu has vivid childhood memories of working in the paddy fields, which involved clearing the woods to make them suitable for farming. Tactile and olfactory images define the relationship between the slush of the fields and the Adivasi worker. Janu recalls that "while working in the fields all our people would smell of the slush day in and day out" (13). This intimate sense of belonging to land is what Peter Berg and Raymond Dasmann refer to as "living-in-place", which involves "following the necessities and pleasures of life, as they are uniquely presented by a particular site" (1977, 399).

When Janu describes the torching of the *punam* (forest) and slashing of the undergrowth, the language comes alive with metaphors connecting nature and the feminine. For instance, a hill that catches fire and blazes for several days is compared to "a woman with her hair shorn, the wild water all blood-red gushing angrily" (Bhaskaran 2004, 2). The forest is also presented as a great nurturing feminine presence and Janu is emphatic in pointing out that "in the forests one never knew what hunger was" (ibid.). She recollects images of plenty from her childhood when the tribal children would catch fish from little streams and ferret out crabs hiding in the slush of fields or feed on the abundantly growing yam. A simple dinner consisting of the cooked fish and crabs gathered in the fields marked the end of the day's labour. The unpierced darkness of the night in the forests became an opportunity for the community to gather in an open courtyard lit by a single ember: "We would sit for hours listening to what the forests mumbled" (3). Mainstream cultures assume that nature is silent, whereas animistic cultures are attuned to the voice of nature expressed by the "wind, earthworms, wolves and waterfalls – a world of autonomous speakers whose intents [...] one ignores at one's peril" (Manes 1996, 15).

For the tribals, survival is predicated on understanding the language of the forest. Perched on a makeshift shack, an *erumaadam*, built between giant trees, the tribals keenly watch the strong gusts of wind, which signal the coming of rain. Janu reminisces about the sense of awe they would feel on seeing big trees swaying in the wind. They would wonder how the bird nests and beehives that hung on the swaying trees manage to keep their place. They would listen with attention to the sounds made by the herds of approaching elephants. They had to remember to run downhill and never uphill if confronted by an elephant to avoid the risk of being overtaken by the animal. They learnt to track elephants by their spoor and chase them away by lighting a fire. They could predict the direction of the winds and the coming of the rains. They lived in harmony with the changing rhythms of nature. As Janu puts it: "We created a system of life for ourselves through centuries of direct observation of earth and Nature" (Bhaskaran 2004, 47).

The second part of the narrative is a lament about the transformation of "Mother Forest" into a "Departmental Forest" with "barbed wire fences and guards" (ibid, 30). This section of the narrative details the encroachment of the land by outsiders who had no knowledge of working with it and whose sole objective was "to extract profit, instead of yield, from the land" (ibid.). The objective of Adivasi evictions in the 70s and 80s was also interconnected with ecological imperialism (Crosby 1986), which was fuelled by the developmental desire to replace the biodiversity of the forests with a eucalyptus monoculture and carve out a wildlife sanctuary in tribal living spaces. Yi-Fu Tuan points out that in traditional societies, the native's relationship with place is complex and is expressed "through behaviour, local tradition, lore and myth" (1990, 63). The same argument is made by native American authors, who emphasize that an indigenous relationship with land is significantly different from a non-native's understanding (Schweninger 2008, 2). Janu's narrative also points to the outsider's disregard for the tribal knowledge of the forest. The forest produce gathered by the Adivasis, which was a valuable commodity in the barter economy, was banned by those who Janu refers to as "people from the civil society", who claimed that tubers from the forest are poisonous and that the Adivasis would be better off buying potatoes and grain from the local shops. These new practices and farming methods triggered a vicious cycle of poverty and lack that Janu traces: "There is no money. And there is no work. There is no work on the land. Land is a commodity for sale and other deals. Lands are left fallow because they claim that agriculture is not profitable" (Bhaskaran 2004, 52).

The repercussions of the loss of land and labour are multipronged. The Adivasi children became frightened of the forest that can no longer accommodate them or satisfy their needs and the women were transformed into a sexual commodity by the Party men for whom they "bore babies for a pinch of tobacco or a stone necklace or some food" (35). However, Janu finds a strong vein of resilience in the Adivasi women, who did not adopt the ways of the mainstream society – unlike their men, who were seduced by alcohol and tobacco. This capacity to endure also springs from their deep-rooted engagement with the land. "Theirs is a resolve that is hardened by the wind and the rain of the forest and in the face of other difficulties" (53). Janu's narrative also exposes the complexities of the politicization of the Adivasi condition in which the tribals are reduced to a mere vote bank. Janu's exit from the Communist Party[3] was based on the realization that "it [had become] one of the needs of the Party to keep this community poor, starved and incapable of resistance" (40). However, Janu's tone is upbeat and insurgent despite all these odds, and like the legendary Antaeus, she draws her strength from her land and forests: "As long as the earth, the forests, the sky and the wind exist new lives come into being, basking in Nature's sunshine, drawing energy from the winds and continuing to exist, glowing and resurgent" (30).

C. K. Janu's life narrative is an evocative example of what Yi-Fu Tuan calls topophilia, "the human being's affective ties with the material environment", which may manifest as a tactile response and "delight in the feel of air, water, earth" and more permanently as a feeling "towards a place because it is home, the locus of memories, and the means of gaining a livelihood" (1990, 93). The re-enchantment of the commons that Janu advocates through her activism and life work emerges from the consciousness of a woman who has become the forest. Janu's life work resonates with that of John Seed, the Australian rainforest activist. When asked by a reporter why he chose to do earth work despite the struggles he faced resisting lumber companies and politicians so as to save the remaining forests, Seed replied, "I am a part of the rainforest protecting itself" (Macy 2009, 239). In Janu's case, the forest is the maternal earth body, which provides sustenance to her and the community; protecting the forest therefore necessarily becomes an act of self-protection. It is the awareness of this deep emplacement that informs the lyrical pastoralism of the first half of Janu's life narrative and the environmental justice activism of the latter half. However, both parts of this life narrative offer multiple relational frames to de-centre familiar ways of knowing and resituate marginalized stories as embodied experiences of the histories of place and its people. The narrative's re-enchantment strategy lies in resuscitating what Val Plumwood (2008) refers to as "sacrificial and shadow places" by exposing ecological entanglements with local power structures. This leads to pertinent questions about the politics of dwelling when forests become "governmentalized" – questions about whose dwelling place is sacrificed and how this process is obscured and rendered unaccountable when affective places undergo erasure and are forced to become places of production.

Woman as Tree: Sumana Roy's Vision of a *Sahrydaya* Planthroposcene

How I Became a Tree makes a forceful case for the "vegetal turn in Humanities" (Woodward and Lemmer 2019, 23), plant personhood and "non-neurological modes of thought" (Delistraty 2019) that demonstrate an evolved form of intelligence. This text powerfully articulates the idea of the "Planthroposcene", a term coined by Natasha Myers that debunks the singularity of human agency in Anthropocene thinking and replaces it with "projects of co-becoming" in the Planthroposcene in which people become aware of their "profound interimplication with plants" (2017, 297–299). Myer's "co-becoming" resonates with Roy's invocation of the Sanskrit word *sahrydaya*, the "co-soul, the soul mate, the sharer of the soul" (2017, 39), which becomes a marker of the camaraderie shared by people who regard the "tree as a human or the human as a tree" (30).

How I Became a Tree is a multistranded non-fiction work, which knits together fields and genres as diverse as literary history, nature studies, spiritual philosophy, myth, memoir and botanical research. The author is drawn to the idea of becoming a tree because of its non-violent ways of inhabiting the earth and creates an identity-forming, bio-centric narrative collage that "allows new narratives to emerge through the complex interdependencies between plants and humans" and has "implications for ethics, politics and sustainability" (Gagliano, Ryan and Vieira 2017, 14). The narrative meditates on themes such as "women as trees", "plants as children", "forest texts of India" and mythical tales of humans transforming into trees, and thereby addresses the "ecological unconscious" (Roszak 1992) of a nation where reducing nature and women to objects of consumption has become a way of life. Roy's whimsical question "Do men kill trees the way they kill human beings?" captures the intensity of ecological violence in the country, which is caused by uneven development plans (2017, 218).

Roy's work makes a case for affective ecology in an attempt to make a radical departure from the dizzy speed of a machine-dominated world and enter into a state of "tree time", which she defines as an intense dwelling in the present moment, "a life without worries for the future or regret for the past. There's sunlight: gulp, swallow, eat, there's night: rest" (6). However, this "tree time" discourse is not merely a philosophical embrace of the "now" but also a passionate enquiry into the age-old literary conventions that equate women with flowers and with *prakriti*, the dynamic force of nature. Roy draws attention to a grandmother's tale from Bengal that equates a princess with flowers and a nescient slumber that requires the intervention of a chivalrous male to restore the stagnant life world of the sleeping princess to life, as a consequence of which fruits and flowers blossom in profusion. The princess covered by flowers in the folktale is compared to other feminine presences in folktales across India, such as "The Flowering Tree" (translated by A. K. Ramanujan). The woman in this folktale is desired for her capacity to turn into a flowering tree and for the same reason is mutilated and reduced to a misshapen thing until the reviving touch of her husband restores her to her former glory. Commenting on the ravaging of woman and nature in this folktale, Ramanujan writes,

> It becomes almost a sexual ritual, a display of her spectacular talent to turn him on [...] when she is returned to her human state, she too is left, ravaged, mutilated. It is a progressive series of violations till she finally ends up being a "thing" [...] When is a woman safe in such a society? She is most vulnerable when she is a tree.
>
> (as quoted in Roy 2017, 116)

Roy juxtaposes this folktale with the fate of women in the country who are subjected to honour killings for having married against the wishes of their family. She critiques the social construction of passivity that is associated with both women and nature, and catalogues a list of Indian flora like the *parijat* and *rajnigandha* whose names have been bestowed on women, noting that "beauty, fragrance, colour, softness, the ease with which it can be plucked, displaced from plant to vase [...] all these together have turned the woman into a flower in the male gaze" (2017, 8). The author rebels against this false dichotomy in which the tree or plant, which works all day for food, is perceived as the patriarch and the flower as the passive feminine. To mark her protest, Roy foregrounds the "tree-as-woman" genre of writing, citing Bangla writers such as Balai Chand Mukhopadhyay, whose short story "The Tree" closely juxtaposes a neem tree and a housewife's attraction for a man who admires the tree for its innate beauty and not for any instrumental value. Roy also invokes Greek and Roman myths of humans and divine beings who were transformed into trees due to fear of sexual violence. She observes that although contemporary transformation narratives from India do not involve trees, almost all Indian folktales have "at least one tale where a girl, harassed, killed, or volitionally, [...] [is] turned into a tree" (21). She cites the "Pomegranate Queen Tale", which is mentioned in Ramanujan's collection of folktales. The folktale narrates the cruel murder of a daughter by her father after she refuses to get married to a man of his choice. However, her chopped-up body, which is buried in the garden, enjoys a vegetal resurrection in the form of a beautiful pomegranate tree. A variant of this story can be found in folktales from all over India, indicating the basic rubric of sexual violence that informs the lives of the women from all parts of the country. Roy connects these myths to the chilling gang rape and torture of a 33-year-old physiotherapy intern, internationally known by the pseudonym Nirbhaya, in the Indian capital, New Delhi, in 2012.

Roy also highlights a vegetal subaltern politics that lies at the heart of the invisibilization of plant life from the framework of history and the idea of an "eventful" *bios* that is restricted only to human beings.[4] In the process of examining the mechanisms that separate the silent language of plants from human speech and arguing for a connection that goes beyond the language of exclusion, Roy observes that "our dismissal of plant life comes from our inability to engage with nothing happening" and notes that this has to do with the privileging of beings with the faculty of sight over those that lack it (15). As a result, the invisible workings of plant life go largely unnoticed while a tree's health and importance are judged solely based on its new flowers, fruits and leaves, which have instrumental value to humans. Roy therefore concludes that "secrets have never been given space or even footnoted in any history book, and so plants continue to remain ahistorical creatures" (16). In contrast to the image of the tree-as-weakness embodied in these narratives of rape, Roy also offers images of the tree-as-body as a symbol of strength that informs Indian cultural life. She refers to the plant-as-person equivalence held in the moral vision of rulers like Vansh Pradip Singh, the ruler of the kingdom of Sawar in northern India in the early twentieth century, who declared, "If you cut the smallest branch of a tree it is just as if you cut my finger" (23). Monica Gagliano refers to this relationship between humans and place as "an equivocal one of respect" characteristic of most indigenous cultures, which "accepts all beings 'as is' in a non-controlling and non-hierarchical way", thereby enabling the entry into "a level of entangled reality where there is no time and no separation into self and other – hence no conflict, no destruction, no ecological crisis" (2018, 191). Roy explores the concept of entangled reality through the sense of touch by reference to the Chipko Andolan, a Gandhian forest conservation movement led by peasant women in the Garhwal Himalayas in 1973, who protested against the rampant deforestation of the region that deprived them of firewood, fodder for livestock and water for drinking and irrigation. They hugged trees to prevent tree loggers and very soon the movement became a *tour de force* that inspired other environmental movements in India. Roy ends her reflection on how "trees became people" by giving a Global North perspective on people forming affective bonds with trees in Melbourne. Civic authorities in the city gave trees email addresses with the objective of getting the public to report on problems like broken branches. However, what transpired instead was the strange case of people emailing their favourite trees, "confiding secrets and insecurities, asking for counsel and compassion" (2017, 27). Roy concludes that in the process of becoming "letter boxes and shrinks", trees attain personhood. The ecofeminist thrust of movements like the Chipko Andolan that are prompted by typical Global South scenarios, where abuse of nature results in the decline of women, and the post-consumerist Global North scenarios, where people form affective bonds with trees due to a lack of communal bonds with other humans, are both indicators of our vegetal entanglements, which are necessary for our material and psychological well-being. Roy cites the example of the artist Nandalal Bose, whose paintings and essays on drawing trees offer a perspective on a special kinship with plants that places them on the same level as humans and animals. This "plant-as person" equivalence becomes a gestalt worldview of sorts for Bose, who sees the form of trees in everything. In his collection of essays on how to draw trees, Bose says that he sees trees in mountains and that if one has mastered the art of drawing plants, one will know how to draw mountains too. Roy's deep engagement with plant life results in a kind of writing that she describes as "therapeutic". Drawing inspiration from the tree photographer Berth Moon, who refers to the thousand-year-old trees that she photographs as "monuments" and her photographs as "portraits", a genre normally reserved for humans, Roy ponders whether it is possible to write like a tree. Tree time with its "expansive caesuras" would inspire a project not bound by the rush of linear time but rather an elaborately slow work of creativity, "a something I didn't want ever to end" (62). Roy's dispersed meditations concerning the understanding of plants as integral actors in the health of the planet as well as the cultural ecosystem constitute what Giovanni Aloi refers to as "the recuperation of 'alternative gazes' and the crafting of new ones – modalities of engagement [...] that can lead to a reontologization of the living" (2018, xx). Roy hails Santiniketan, the sylvan retreat of learning developed by Tagore and his family, as the exemplar of this paradigm shift. In a letter to his writer friend, Kshitimohan Sen, Tagore reveals that the ashram's calendar was based on seasons and the flowering of plant life: "The summer-aided forest festival, the saptaparni-chhatim monsoon festival, the shefali festival in autumn, the spring festival in the sal forests" (Roy 2017, 97). Tagore's school in the woods taught the children to relate to trees as friends, participants and neighbours in their daily lifescapes and not as mere props in the background. According personhood to trees came naturally to Tagore. On trips away from Santiniketan, he would write letters enquiring about the health of an ailing creeper, expressing his longing for a favourite pine tree or agonizing about drawing up the right watering schedule for the garden. Roy draws attention to Tagore's habit of baptizing his plants with new names for the sheer "joy of naming an unfamiliar flower" (90). Although Roy initially finds names, as an indicator of identity, rather limiting, she also has a moment of deep insight and rapture when a "foreign" tree planted in the courtyard of her house begins to "reveal its identity through clusters of blue flowers at every node" (92). Propelled by a desire to speak as a poet to the tree whose blue flowers instilled silence and happiness in her, Roy decides to name the creeper *neelamonilata* or the blue iris creeper.

Writing about a "plant-like" notion of subjectivity that would dissolve the rigid boundaries between nature and culture, Elaine Miller proposes that

> One may start with an idea, or start with a very straightforward reading of a text. Then the "seed" metamorphoses into a "stem", "node" or "leaf" and without any specific intention on the part of the author, it begins to transform itself into something else [...] One's final reading will never be final, never exhaustive.
>
> (Miller 2002, 11)

This open-endedness can be witnessed in Tagore's short story "Bawai", in which a silk cotton tree is the child narrator's doppelgänger. Roy draws attention to Tagore's child protagonist, who is described almost entirely in vegetal terms:

> When it rained, he behaved like a happy tree in July, when it grew very hot, he moved about bare-chested embracing the heat like trees do. He was always tree-like in the month of Magh, he behaved like the mango tree beginning to bloom; in Phalguni, he became a flowering sal tree.
>
> (Roy 2017, 101)

Tagore's tree poems for children establish a special kinship with trees. For instance, banyan trees are referred to as patriarchs, whose branches satisfy the child's desire to play and swing on them "like a grandchild does in a grandparent's lap" (99). Roy notes that in these poems there are both human and vegetal actors, and no metamorphosis but rather a "beautiful fluidity, where one could be both, depending on desire and circumstances" (ibid.).[5]

The forest as a site for re-enchantment also figures prominently in Roy's work. The forest is described both as a material reality and a state of mind. Recalling her experience in the Sukna Forest in the eastern Himalayas, where she felt a delirious happiness and also a strange guilt for trespassing into the forest, Roy declares, "The forest needs me as much as I need the forest. The symbiotic nature of this need took me years to comprehend" (160). These dualisms merge and dissolve at a more intense level of "forest bathing": "It is eating soil like an earthworm so that the earthworm also becomes the earth" (164). The contingency and uncertainty contribute to the terrible beauty of the forest, which Roy defines as "the lack of superintendence" and "the beauty of being temporarily abandoned" (164). She describes sojourning in a forest as an antiquated form of time travel, since the traveller is privy to "accretions of unnoticed history made visible in stumps and trunks of trees" (159). The protagonist of Bibhutibhushan Bandopadhyay's novel *Aranyak* (1976) is an estate manager who goes to the jungles of Bihar to distribute land and clear the forest. He is so enchanted by the timelessness of the forest that he "can see a mythico-historical figure like Krishna or imagine the river Yamuna in the middle of a severe drought" (159). The forest, therefore, becomes a state of mind for the enchanted or, as Thomas Berry puts it, an expression of "our intimacy with the universe", which when developed to its fullest potential results in the forest and fauna presenting themselves "to our opening senses. Each in its own distinctive perfection fills our mind, our imagination, our emotional attraction" (Berry 2006, 5–6). Commenting on how the Buddha's life was intertwined with plants, Roy observes that venerating a Bodhi Tree with humility, as if one were in the presence of the Buddha himself, is an example of the metonymic relationship between the Bodhi Tree and the Buddha. "I grew certain that the Buddha was a tree, an evergreen one" (Roy 2017, 197). Religious discourses that hint at these equivalences advocate the enchantment of nature through the reawakening of the sense of awe and wonder regarding the source of creation, a point eloquently argued by Alister McGrath in his book *The Reenchantment of Nature* (2002), where he reclaims the idea of nature as God's creation. The re-enchantment principle helps Roy to theorize death as an eternal "return to life through nature" (221).[6] Roy concludes her arboreal meditations by declaring that if she ever became a tree, she would "choose to be an Ashoka, the a-shoka, the sorrowless tree" (ibid.).

Conclusion

In the Indian cultural context, the forest is associated with *vanaprastha*, the forest-dweller stage in life, which is the prelude to *sanyasa* or a life of renunciation. However, the forest is never a space of perfect democracy, and both C. K. Janu and Sumana Roy focus on the tragedy of the commons that has been a part of India's environmental history. For instance, epics like the Mahabharata describe the burning of the Khandava Forest and its non-human inhabitants as a necessary act to provide a clearing for the Indraprastha palace, a powerful symbol of urban living. The clearing of the forests and the destruction of the commons are staple features of contemporary India. Conflicting ideas inform Indian forest policies, with debates revolving around whether utilization of forest lands should cater to developmental demands or take into account the needs of the marginalized peasants, pastoralists and tribals whose lives are imbricated with the life of the forests. This "militant ethic of the forest" is reflected in the narratives of writers like Bibhutibhushan Bandyopadhyay and Mahasweta Devi. Drawing on Michael Marder's idea of "a vegetal model of the political", which recognizes the "anarchic proliferation of multiplicities, of branches and twigs that retain their semi-independence while participating in the overall growth of plant society" (as quoted in Roy 2017, 166), Roy draws an analogy between individual trees that refuse to see themselves as part of a collective and the "indifference of forest dwellers to see themselves as part of a nation" (Roy 2017, 166), thereby bringing into focus the highly contested forest spaces in India that have been reclaimed for settlement and agriculture since colonial times. Post-independence, the enforcement of enclosures and developmental projects like dams, mines and roads has mostly affected the Adivasis and rural poor. Roy discusses "Arjun", a popular short story by Mahasweta Devi, to bring home the stark reality of the environmentalism of the poor (Guha and Martínez Alier 1997) in a tribal village where illegal felling of tree fuels the timber economy. A character in the story sums up the situation: "At the end of the day I need some money. If you tell me to cut a tree, I'll cut a tree, if you ask me to cut a man, I'll cut a man" (219). However, the hoary Arjun Tree of the village is saved by an act of sacralization – it is declared a village deity and worshipped with drums and music, which keeps away the loggers at least for the time being.

Roy's narrative strategies bring to centre stage a range of issues that adversely affect India's forest communities, from myopic development plans to mass clearing of trees, thereby pointing to a process of silent erasure that has accelerated the Adivasis of India to the same tipping point as the native Americans and the Australian aborigines. The *sahrydaya* co-souling ethics espoused by both these texts points to the way in which women writers from the Global South forge new ways of re-enchanting the earth. C. K. Janu's life narrative details the unity in tribal culture, which is impervious to the homogenizing ways of mainstream society, and attributes this resilience to the tribal woman, who "is hardened by the wind and the rain of the forest and in the face of other difficulties" (2004, 53) and draws her strength from the elemental forces of nature. In a similar fashion, Sumana Roy's narrative draws attention to how the "woman of the forest is a figure so enchanting" that she can be seen both as a figure of repose and "a powerful symbol of civilization of the forest" (2017, 161), who challenges the urban imagination that seeks to tame and instrumentalize her. The different facets of the arboreal feminine foregrounded in these narratives echo the ancient forest culture of India, which vested the power of protecting forests in female deities like Bonbibi and Aranyani. By emphasizing that humans share their lifeworlds with vegetal non-human nature, these texts signal the need to exit the Anthropocene and enter the Symbiocene – a term used by Glenn

Albrecht (2019) as an alternative to the Anthropocene, with all its negative connotations of disenchantment and apocalyptic espousals of ecosystems that have gone beyond repair.

By creating new alter-tales of human kinship with more-than-human others, these texts challenge the "narratives of disenchantment" that construct the "everyday world as banal" (Bennett 2001, 80) and enable the reader to be receptive to the unprecedented manner in which "things hail us" (Bennett 2010) in the everyday world. The recognition of this vibrant agentic power inherent in vegetal life becomes a spur both to creative activity and environmental justice activism.

Notes

1 The *oikeion* bleeding into the *politikon* is a unique aspect of Global South ecofeminism. Interrogating the Greek concept of the *oikeion* as the space for women, servants and children and the oppositional *politikon* as the male, public space, Lyotard points out that the *oikeion* is the private space outside the public sphere, the "shadowy space of all that escapes the light of public speech" where "the tragedy occurs" (2000, 135). Women in the Global South who care for nature either through their writings or activism have reinterpreted the *oikeion* as a site of struggle and self-actualization by interweaving the private and the public spheres. Their writings and life narratives embody a new materialism, since they emphasize notions of interconnection and entanglement and a non-hierarchical relationship in the natureculture continuum (Latour 1993), which has strong ethical and political implications. This paradigm, as Christa Grewe-Volpp observes, "erases notions of subjectivity that rely on human exceptionality. It fosters instead a more humble recognition of humans being part of an ever evolving world" (2016, 218). According to Carolyn Merchant, women deeply engage with earthcare of their oikic spaces when "their bodies, or the bodies of those with whom they have a caring relationship, are threatened by toxic or radioactive substances or when land or another species about which they care deeply is threatened with extinction" (2016, xviii). Notable examples from India include Amrita Devi and her band of 363 ecowarriors from the Bishnoi community of Rajasthan, who in 1730 sacrificed their lives to protect the *khejri* forest from the felling operation ordered by the king of Jodhpur, and the tree-hugging women protestors of the non-violent Chipko movement that commenced in 1973.

2 A trans-scalar perspective involves zooming both in and out in order to get the finer details as well as the big picture. It fosters the ability to forge connections between the local and the global and to imagine lived experiences and issues concerning gender and ecology across scales, which, in turn, leads to a better understanding, informed praxis and holistic ways of being in the world.

3 Following its victory in the state of Kerala's first democratic elections in 1957, the Communist Party became a force to be reckoned with in Left politics in Kerala and received overwhelming support from Scheduled Caste and Scheduled Tribe voters. However, things changed in the 1990s with the emergence of the Adivasi rights movement and the formation of the AGMS (Adivasi Gothra Maha Sabha) in 2001, which organized social movements that agitated for the redistribution of land to landless tribal people in Kerala. C. K. Janu joined the Communist Party of India as a literacy campaigner. However, she quit the party as a result of her disillusionment with the party's patriarchal, anti-tribal policies and became the leader of the AGMS, which carved out an independent trajectory of protest for the Adivasis.

4 Roy's woman–nature juxtapositions resonate with the core vision of Elaine Miller's work *Vegetative Soul* (2002), in which Miller argues that the vegetal can be a powerful site for reconfiguring the female subject, since the feminine, like the vegetal, has been devalued and is considered the passive partner, whereas the male animal is coded as the active part of the binary. According to Miller, "a change in the symbolics of nature necessarily brings about a change in that of culture" and therefore "the possibility of restructuring or even creating feminine subjectivity in a way that would make a real difference to women" (2002, 4).

5 Roy also foregrounds the contribution made by Tagore's great contemporary botanist Jagadish Chandra Bose, who experimented on what he referred to in Bangla as *torulipi* – the "script of plants", the plant autographs that revealed the "secret life of plants". In one of his public lectures, Bose spoke about the longstanding error in observation and theory that categorized the vegetal as insensitive. Referring to the spectacular performance of the Praying Palm Tree of Faridpore, which bows down as if prostrating itself every evening, Bose drew attention to the tremendous capacity of plants to be fully alive to changes in their environment. In an essay titled "Nirbak Jeebon" or "The Silent Life", Bose sought to discover the "direct speech" of plants unmediated by human language and intentionality, and subjects them to an instrument that can communicate a plant's response to pain stimuli. Roy has a moral objection to the prison-like torture that the subjects of Bose's investigation had to undergo in order to reveal their secrets to him. Bose's mission of mapping the history of torture by pinching and breaking his plants and recording their reaction on different instruments was based on his scientific definition of life as a living creature's immediate response to pain. Roy's critique is an invocation of the "ontopoetics" of nature, to use ecophilosopher Freya Mathew's term. According to Mathews,

> nature must be understood by us as structured by meanings as well as by physical causality: reality must be recognised as a psychophysical system with an inner, poetic aspect as well as an outer, causal one, capable of capturing and commanding us emotionally as well as making a claim on our moral consideration.
>
> (Matthews 2010, 8)

6 The perfect symbol of this positive transformation is the Bios urn (Capsula Mundi) designed by Anna Citelli and Raoul Bretzel, which is divided into two capsules – with the upper half containing soil and seeds and the lower one the ashes of the dead.

References

Albrecht, Glenn. 2019. *Earth Emotions: New Words for a New World.* Ithaca, NY: Cornell University Press.

Aloi, Giovanni. 2018. *Why Look at Plants? The Botanical Emergence in Contemporary Art.* Critical Plant Studies: vol. 5. Leiden: Brill.

Bennett, Jane. 2001. *The Enchantment of Modern Life: Attachments, Crossings and Ethics.* Princeton, NJ: Princeton University Press.

Bennett, Jane. 2010. *Vibrant Matter: A Political Ecology of Things.* Durham, NC: Duke University Press.

Berg, Peter, and Raymond Dasmann. 1977. "Reinhabiting California." *Ecologist* 7 (10): 399–401.

Berry, Thomas. 2006. "Prologue: Loneliness and Presence." In *A Communion of Subjects: Animals in Religion, Science, & Ethics*, edited by Paul Waldau and Kimberley Patton, 5–6. New York: Columbia University Press.

Bhaskaran. 2004. *Mother Forest: The Unfinished Story of C. K. Janu.* New Delhi: Kali for Women.

Bijoy, C. R., and K. Ravi Raman. 2003. "Muthanga: The Real Story: Adivasi Movement to Recover Land." *Economic and Political Weekly* 38 (20): 1975–1982. https://www.jstor.org/stable/4413574.

Crosby, Alfred W. 1986. *Ecological Imperialism: The Biological Expansion of Europe, 900–1900.* Cambridge: Cambridge University Press.

Delistraty, Cody. 2019. "The Intelligence of Plants." *Paris Review,* September 26. https://www.the parisreview.org/blog/2019/09/26/the-intelligence-of-plants.

Gagliano, Monica. 2018. "Persons as Plants: Ecopsychology and the Return to the Dream of Nature." In *Covert Plants: Vegetal Consciousness and Agency in an Anthropocentric World,* edited by Prudence Gibson and Baylee Brits, 183–193. Santa Barbara: Brainstorm Books.

Gagliano, Monica, John C. Ryan and Patricia Vieira, eds. 2017. *The Language of Plants: Science, Philosophy, Literature.* Minneapolis: University of Minnesota Press.

Gibson, James William. 2009. *A Reenchanted World: The Quest for a New Kinship with Nature.* New York: Holt.

Grewe-Volpp, Christa. 2016. "Ecofeminisms, the Toxic Body, and Linda Hogan's *Power.*" In *Handbook of Ecocriticism and Cultural Ecology*, edited by Hubert Zapf, 208–225. Berlin: De Gruyter.

Guha, Ramachandra, and Joan Martínez Alier. 1997. *Varieties of Environmentalism: Essays North and South.* London: Earthscan.

Kozhisseri, Deepa. 2020. "Valli at the border: Adivasi women de-link from settler colonialism paving re-enchantment of the forest commons." *Journal of International Women's Studies* 21 (7): 139–152. https://vc.bridgew.edu/jiws/vol21/iss7/11.

Krishna, Nanditha. 2020. "Ancient Forests and Sacred Groves." In *Critical Themes in Environmental History of India*, edited by Ranjan Chakrabarti, 168–209. New Delhi: Sage.

Kurth-Schai, Ruthanne. 2005. "Ecology and Equity: Toward the Rational Reenchantment of Schools and Society." *Educational Theory* 42 (2): 147–163. http://dx.doi.org/10.1111/j.1741-5446.1992.00147.x.

Latour, Bruno. 1993. *We Have Never Been Modern.* Cambridge, MA: Harvard University Press.

Linebaugh, Peter. 2019. Foreword to *Re-enchanting the World: Feminism and the Politics of the Commons* by Silvia Federici, xiii–xvii. Oakland: Kairos.

Lyotard, Jean-François. 2000. "Ecology as Discourse of the Secluded." In *The Green Studies Reader: From Romanticism to Ecocriticism*, edited by Laurence Coupe, 135–138. London: Routledge.

Macy, Joanna. 2009. "The Greening of the Self." In *Ecotherapy: Healing with Nature in Mind,* edited by Linda Buzzell and Craig Chalquist, 238–245. San Francisco: Sierra Club Books.

Manes, Christopher. 1996. "Nature and Silence." In *The Ecocriticism Reader: Landmarks in Literary Ecology*, edited by Cheryll Glotfelty and Harold Fromm. Athens: University of Georgia Press.

Mathews, Freya. 2010. "On Desiring Nature." *Indian Journal of Ecocriticism* 3: 1–8.

McGrath, Alister. 2002. *The Reenchantment of Nature: The Denial of Religion and the Ecological Crisis.* New York: Doubleday.

Merchant, Carolyn. 1996. *Earthcare: Women and the Environment.* New York: Routledge.

Miller, Elaine P. 2002. *The Vegetative Soul: From Philosophy of Nature to Subjectivity in the Feminine.* Albany: SUNY Press.

Morton, Timothy. 2016. *Dark Ecology.* New York: Columbia University Press.

Myers, Natasha. 2017. "From the Anthropocene to the Planthroposcene: Designing Gardens for Plant/People Involution." *History and Anthropology* 28 (3): 297–301. http://dx.doi.org/10.1080/02757206.2017.1289934.

Naess, Arne. 1973. "The Shallow and the Deep, Long-Range Ecology Movements." *Inquiry* 16: 95–100.

Plumwood, Val. 2008. "Shadow Places and the Politics of Place." *Australian Humanities Review* 44: 139–150. https://pressfiles.anu.edu.au/downloads/press/p38451/html/shadow.xhtml?referer=&page=14.

Raman, Ravi K. 2002. "Breaking New Ground: Adivasi Land Struggle in Kerala." *Economic and Political Weekly* 37 (10): 916–919.

Rangarajan, Swarnalatha. 2016. "Women Writing Nature in the Global South: New Forest Texts from Fractured Indian Forests." In *Handbook of Ecocriticism and Cultural Ecology*, edited by Hubert Zapf, 438–458. Berlin: De Gruyter.

Rose, Deborah Bird. 2004. *Reports from a Wild Country: Ethics for Decolonisation.* Sydney: UNSW Press.

Roszak, Theodore. 1992. *The Voice of the Earth.* New York: Simon & Schuster.

Roy, Sumana. 2017. *How I Became a Tree.* New Delhi: Aleph.

Schroeder, Ralph. 1995. "Disenchantment and its Discontents: Weberian Perspectives on Science and Technology." *Sociological Review* 43: 227–250.

Schweninger, Lee. 2008. *Listening to the Land: Native American Literary Responses to the Landscape.* Athens: University of Georgia Press.

Shiva, Vandana. 1989. *Staying Alive: Women, Ecology and Development.* London: Zed Books.

Tuan, Yi-Fu. 1990. *Topophilia: A Study of Environmental Perception, Attitudes and Values.* New York: Columbia University Press.

Woodward, Wendy, and Erika Lemmer. 2019. "Introduction: Critical Plant Studies." *Journal of Literary Studies* 35 (4): 23–27. http://dx.doi.org/10.1080/02564718.2019.1690804.

Part 2

Materials

Agency in/of Transcorporeal Assemblages

4 Plastic Ambivalence

Nicole Seymour

In late 2017, a massive backlash against glitter erupted across US and UK media. It was sparked in large part by New Zealand-based environmental anthropologist/political ecologist Trisia Farrelly's declaration to *The Independent*: "'I think all glitter should be banned, because it's microplastic'" (Gabbatiss 2017) – which means it passes through filtration mechanisms, flows to the ocean, and potentially ends up being ingested by sea life, all the while leaching toxic chemicals. Largely ignoring the fact that there are actually many non-plastic, biodegradable glitter alternatives, various outlets prepared their clickbait. CNN soon reported that "Glitter is Not Just Annoying, It Could Be Bad for the Environment" (Willingham 2017), while a not-so-subtle *Newsweek* headline blared, "Glitter Is an Environmental Scourge That Is Wrecking the Oceans. Should It Be Banned?" (Sheridan 2017).

Plastic pollution is a significant issue that has long concerned scientists and activists, with an estimated 12.7 tons of plastic entering the oceans each year ("Plastic Pollution" n.d.). But in this case, the media distorted the issue; glitter is not a significant source of microplastics compared, for example, to fibres from fleece and other synthetic clothing.[1] I would also note that this media backlash fits mainstream environmentalism's typical "doom and gloom" script, in which everything – even the most seemingly innocent, fun and miniscule objects or behaviours – are "bad for the environment." Such scripts, as I argue in my recent book *Bad Environmentalism: Irony and Irreverence in the Ecological Age* (2018), leave humans feeling guilty, ashamed and overwhelmed. Further, they prevent us from thinking in nuanced and complex ways about our extensive entanglements with environment and about larger structural forces – which, conveniently, get obscured in the emphasis on individual behaviours.

Plastics discourse is thus clearly an issue of affect, or emotion, regarding our enmeshment in an environmental condition. But beyond the affective or emotional dynamic of the media's appeals and ensuing public reactions, I am interested in humans' affective relationships to plastic materials and commodities themselves. Indeed, the glitter backlash unleashed passionately polarized feelings, sometimes even from the same person. As blogger

DOI: 10.4324/9781003023319-7

Penny Whitehouse observes: "Glitter. You either love it or hate it, right? I love it and hate it. Is that even possible? I love it because it's pretty but I hate it because it's bad for the environment" (n.d.). As this comment suggests, human emotions around plastic materials and commodities are deeply mixed.

Ambivalence about plastics is certainly nothing new. In fact, cultural researchers have traced it back to at least the 1940s. Jeffrey Meikle's book *American Plastic: A Cultural History* notes that, like atomic energy, mid-century plastics "reflected the ambivalence of a utopian hope afflicted with an undercurrent of dread" (1995, 243). Meikle finds that these materials evoke for consumers both "the permanence of death" and the freedom of malleability, and that this "duality [has run] throughout the American experience of plastic" (9). Similarly, Susan Freinkel's book *Plastic: A Toxic Love Story* "traces the arc of our relationship with plastics, from enraptured embrace to deep disenchantment to the present-day mix of apathy and confusion" (2011, 9); she diagnoses "our feelings toward plastic [as] a complicated mix of dependence and distrust" (8). Importantly, both authors note how plastic consumption has increased over time *alongside* concerns about it.

This essay builds on those observations in particular ways, showing how ambivalence around plastics, and toxicity and waste more generally, has become a significant feature of contemporary environmental humanities scholarship and of contemporary art. Throughout, I seek to highlight the queer and feminist dimensions of that ambivalence. In the first section, I will trace how the academic subfield of queer ecology departs from the adamant opposition to plastics and toxicity espoused by mainstream environmentalism to articulate a (limited and qualified) reclamation or embrace thereof. This ambivalence, as I explain, functions both as a measure against "purity politics" (Ah-King and Hayward 2013, 5) and as a pragmatic recognition of our current moment of crisis – though it certainly poses many risks. In the second section, I will turn to the ever-growing genre of contemporary plastic art, demonstrating how well-known male artists such as Chris Jordan (USA), Mark Dion (USA) and Stuart Haygarth (UK) aestheticize plastic while drawing attention to plastic pollution – an aestheticization that, as I show, carries many risks as well. I conclude this section with an in-depth look at two female artists, Ruth Peché (Spain) and Ifeoma U. Anyaeji (Nigeria/USA), who have received comparatively less attention. Like the aforementioned artists, Peché and Anyaeji evince that ambivalent attraction to/concern about plastic, but they offer quite different work: their use of feminized aesthetic practices offers, I will argue, a commentary on the unevenly distributed violence of plastic pollution.

Part I: Theory: The Ambivalent Embrace of Plastic

A brief sketch of the subfield of queer ecology is in order before proceeding. First, as I have summarized elsewhere, "[q]ueer ecology is a conceptual framework that interrogates the relationship between the queer and the natural, broadly construed," and one of its foundational insights is that those two entities have regularly been opposed in public discourse (Seymour 2020, 108). For example, as Robert Azzarello observes,

> [f]rom a queer perspective, what is most troubling in [mainstream environmentalist] discourse is that the most popular terms to describe environmental crisis – unnatural, diseased, pathological, risky, contaminated, suicidal, and so on – are exactly those terms that have been used historically to stigmatize sexual misfits.
>
> (Azzarello 2012, 125)

Considering that plastic products such as glitter have historically been associated with queer communities – as Katie Schaag reminds us, "[q]ueer aesthetics, from camp cinema to non-binary fashion, rely upon plastic materials (glitter, silicone, spandex) as well as plastic-inflected visuals" (2018) – this point prompts me to return briefly to the viral glitter-is-plastic-and-therefore-bad story. While that story seemed to trade on shock value, on the incommensurability of item and impact, it also evokes a more familiar narrative in which queerness, or that which is associated with it, is framed as toxic. That is, the demonization of glitter arguably plays into negative conceptions of sexual minorities, in the name of environmentalism – thereby creating a false dichotomy: queerness *or* environmentalism.

Not all queer ecologists agree on how to respond to such a status quo. While some have tried to recuperate queer connections to nature, others have tried to reframe those links – for example, interrogating what counts as natural to begin with. When it comes to questions of plastic, toxicity and waste more specifically, others have gone a bit further. Mel Chen dwells contrarily on the associations between the category of the toxic and that of the queer, arguing that "queerness and toxicity [...] have an *affinity*. They truck with negativity, marginality, and subject-object confusion; [...] they challenge heteronormative understandings of intimacy. Both have gotten under the skin" (2012, 281). In his own work on glitter as a microplastic, Vanbasten Noronha de Araújo declares, "I think of queers and waste through a shared onto-epistemological disposition to trouble categories, recognizing [their] fluidness" (2019, 6), while Guy Schaffer takes this insight to the disciplinary level, arguing for an affinity between queer theory and what has come to be known as "discard studies"[2] on the grounds that both are

> interested in uneven remainders, things that don't fit nicely into categories. Both concern themselves with the strange and imperfect construction of divisions (in discard studies, that between waste and not-waste; in queer theory, those between hetero/homosexual, between male and female).
>
> (Schaffer n.d.)

To find potential or even pleasure in plastic, toxicity and/or waste may seem to some a rather questionable move, especially from a multispecies justice perspective. To give just two specific examples: plastic bag pollution historically contributed to massive monsoon flooding in Mumbai, India before bans took effect, while "[p]lastic debris has been identified as the cause of injury or death in 267 different species, including [...] 43 percent of all marine animals" (Freinkel 2011, 151, 132). Importantly, then, those queer ecologists acknowledge the potential pitfalls of their arguments. Schaffer, having argued for camp performance as a strategy of both making waste visible and revaluing it, admits that he runs the risk of

> hermeneutical frivolity [by] suggesting that fracking fluid or nuclear fallout might be rendered palatable through camp interpretation[.] [But as Susan] Sontag explains, "camp doesn't reverse things. It doesn't argue that the good is bad, or the bad is good" (61). What a camp lens *can* do is animate the ways in which the disposal of waste is never complete, that the boundaries created between waste and the social worlds that produce it are always partial.
>
> (Schaffer n.d.)

Meanwhile, Chen acknowledges, "I would be foolish to imagine that toxicity stands in for 'utopia' given the explosion of resentful, despairing, painful, screamingly negative affects that surround toxicity. Nevertheless, I am reluctant to deny the *queer productivity* of toxins and toxicity" (2012, 281). And when it comes to plastic, that productivity is wide-ranging. For example, as oceanographers Kara Lavender Law and Richard C. Thompson explain, "floating microplastics in open ocean gyres provide habitats for diverse communities of microorganisms" (2014, 145) – a phenomenon often referred to as the "Plastisphere" – even as they may be contributing to the demise of other species. But perhaps more to the theme of this section, plastic destabilizes sex and gender binaries in both humans and nonhumans through its hormone-disrupting properties, thus producing what theorist Heather Davis has referred to as "queer new worlds." As Davis explains,

> chemicals known as plasticizers are added to make plastic pliable, or hot pink, or heat resistant. Perhaps the most infamous of these is bisphenol A (BPA), one of numerous chemicals known for its reproductive toxicity. BPA literally blocks the human ability to reproduce through both an overexposure to the hormone estrogen and by way of endocrine disruptors that mimic hormones in the body and replace their functions, sometimes queering the gender of the body in which it resides.
>
> (Davis 2016, 189)

Beyond the material production of queer embodiment, Davis is also interested in the "odd ways in which plastic might suggest certain kinds of queer political formation that dovetail with contemporary trends in political organizing" (ibid.) – including proliferation, accumulation and the notion of "nonfilial progeny." In short, plastic is, ironically, *generative and productive* even as it potentially destroys.

This leaves us in a very strange place affectively. We are both horror-struck by the idea of dead marine animals bloated with plastic – famously captured by the likes of photographer Chris Jordan, discussed below – *and* intrigued to hear that life finds a way on life rafts made of plastic. We realize that plastic has revolutionized society and enlivened subcultures and *also* might be the end of society and culture, or, at least, its final calling card. So, what to do? Recognizing this complexity, and thus acknowledging that plastic is many things to many people, is a simple but important response. As Schaag puts it, paradoxically but also pragmatically,

> Studies of the utopian politics of queer aesthetics must consider the environmental cost of materials like glitter and silicone, *and* studies of plastic's toxic impact on the environment should acknowledge the positive role it has played in destabilizing naturalized gender roles and facilitating new forms of pleasure.
>
> (Schaag 2018)

Meanwhile, Davis has argued that,

> Without wanting to celebrate these new toxicities, we must find ways of living with them. [...] To live ethically in the present moment means finding creative strategies for living with toxicity – to accept it as a new queer future and to find ways of navigating horror.
>
> (Davis 2016, 192)

What I find implicit in all of these theorists' work is an affective tempering, in which one refrains from *both* naïve optimism *and* the doom and gloom that seems to dominate environmental reportage, and instead grapples directly with the messy reality in which environmental destruction is already here and, in many cases, irreversible. As Max Liboiron, Manuel Tironi and Nerea Calvillo have pointed out, we inhabit a "permanently polluted world" where all human bodies contain industrial chemicals (2018, 332). More broadly, this tempering shows how far environmental thought has come; the aforementioned theorizing is of a piece with the dawn of the Anthropocene concept, in which humans reject the fantasy of a pure nature "out there" to which we can return, and reject the binaries of nature/culture and human/other that ecocritics and others have been fighting for so long. Like the mid-century consumers of plastic, the aforementioned theorists are ambivalent – but they articulate this ambivalence openly rather than sublimating it, and see it as part and parcel of queer and other politics, not as apoliticism or apathy.

But we must note that this line of thought still poses at least two problems. First, it creates a troubling strange-bedfellows scenario. As Liboiron has put it, referring to plastic's endocrine-disrupting tendencies, "[i]s feminization of male fetuses abnormal, or even pathological? Is it a form of harm? The lesbian, gay, bisexual, transgender and queer (LGBTQ) community has argued that it is not. So, too, has the chemical industry" (2013, 143). This strange-bedfellows scenario is nothing new when it comes to queer thought, a result of its contrary and, arguably, sometimes reactionary, impulses. For instance, queer theorist Tim Dean noted several years ago that the so-called "anti-social thesis" in queer theory – a rejection of the demand for respectability politics, among other things – chimed with "right-wing fantasies about how the 'homosexual agenda' undermines the social fabric" (2006, 826). Thus, it is incumbent upon theorists to consider how, say, embraces of "queer new worlds" created through toxicity reinforce discrimination against queers – as Azzarello reminds us. In other words, how can we acknowledge toxicity and its queer resonance, or reject panic discourse around queer embodiment, without allowing that connection to serve as the grounds for oppression? More specifically, could we begin to *queer* queer identity away from plastic? Second and relatedly, recalling Davis' argument that, "Without wanting to celebrate these new toxicities, we must find ways of living with them," we should ask whom the "we" in question references, and whom it excludes. Black feminist ecocritic Chelsea M. Frazier has pointed out that mainstream environmental discourse "insist[s] on a 'we' and an 'all' that presume a monolithic relationship to and understanding of the ways environmental degradation is experienced" (2018, 169) – often eliding the experiences of people of colour and those in the Global South, queer and otherwise. Thus, theorists must remain cognizant of power differentials in terms of that experience. For whom is it easier to "liv[e] with new toxicities" like plastic pollution? For whom is it harder, and worse?

In what follows, I consider several examples of "plastic art," to use a broad and simplistic term, with a particular eye towards how it either re-entrenches or avoids a monolithic "we." Further, I consider how this art constitutes a direct response to the question of what to do. These artists' response is perhaps more immediate, and certainly more material than the theoretical approaches described above: take plastic and make art with it, imbuing that art with the care and time that disposability culture, curated in large part by plastic manufacturing and consumption, necessarily opposes.

Part II: Art: Ambivalent Curation and Transformation

Archaeologies of the Plasticene: Chris Jordan, Mark Dion and Stuart Haygarth

What art historian and theorist Amanda Boetzkes calls "the widespread trend to visualize waste in contemporary art" – which we can link back to a "long modernist tradition of connecting art to historical detritus" (2019, 2) – includes, not surprisingly, countless examples of art that takes plastic as its primary material. In what follows, I will briefly discuss three well-known "found object" artists and then take a look at two transformational artists who have received less scholarly attention. Across all, I identify the kind of ambivalence described above as well as a sense that the purpose of art is not (simply) to spark action.

Any discussion of plastic art would have to acknowledge the work of US artist Chris Jordan, whose photographs from his 2009 series, *Midway: Message from the Gyre,* have become iconic representations of the Anthropocene or, as it were, the Plasticene. As marine ecologists Linsey E. Haram et al. explain,

> the Plasticene (Age of Plastics) describes the current period in geological history, beginning in the mid-20th century, with the exponential increase in global production, consumption, and disposal of plastic products (Zalasiewicz et al., 2016). The Plasticene concept materialized from research by Corcoran et al. (2014), who suggested that the vast discarding of plastics will be observable in the fossil record, given that plastic can fuse with rock in the process of burning on beaches and rock outcrops to create plastiglomerates.[3]
>
> (Haram 2020, 2–3)

In short, as plastic will outlive us, it serves as a marker of our civilization. One of the most famous of Jordan's photos alludes to this idea; the untitled work gives us a direct view from above of an albatross carcass laying on pebbled ground, its stomach area stuffed with recognizable objects such as a cigarette lighter and plastic bottle caps as well as small shards of broken plastic. This image has been widely circulated across the Internet and social media, sometimes without attribution to Jordan. For example, I recently glimpsed it on Twitter as part of a poster for an environmental studies lecture, as if the image were a simple documentation of a problem to be discussed. One of the most interesting aspects of this circulation, then, is how it allows us to forget that the image is actually work of art – a point to which I'll return shortly.

I had been familiar with Jordan's work for a while when I went to see a recent screening of his documentary film *Albatross* (2017) at my local theatre in Long Beach, California, hosted by Captain Charles Moore, the oceanographer and author of *Plastic Ocean: How a Sea Captain's Chance Discovery Launched a Determined Quest to Save the Oceans* (2011). Filmed on the Midway Islands in the Pacific, the same location for his photo series, *Albatross* documents the births, lives and occasional plastic-related deaths of the titular birds. For example, they may starve or dehydrate because they're too full of plastic, or die after plastic punctures their stomach lining or blocks their oesophagus. Jordan's work captures the aftermath; here,

Susan Freinkel's documentation is evocative, though she is not commenting on Jordan's work directly:

> Every carcass seems a mockery of the natural order: a crumbling bird-shaped basket of bleached bones and feathers filled with a mound of gaily colored lighters and straws and bottle caps. The birds are dissolving back into the ground; the plastics promise to endure for centuries.
>
> (Freinkel 2011, 118)

One could assume that this is the point of Jordan's images: to remind us of the unnatural, enduring dangers of plastic. I therefore found myself intrigued, even surprised, by the fact that artists like Jordan also highlight the beauty and potential of this substance. For instance, when staging a funeral for a dead albatross in his documentary, Jordan ritualistically encircles the bird with colour-aggregated rings of plastic objects (similarly captured in his 2010 photograph, *Albatross Mandala Midway Island*). Here, we can see what others might sometimes ignore about his other artworks, such as the one with which I began: they are art, painstakingly staged and/or framed to be beautiful, even if also horrible. We might also note that Jordan's voice-over diction in the documentary (inadvertently?) invokes the artistic, as when he tells us that "each time I opened [a dead bird] up it was like a *gallery* of [plastic] horrors" (2017). Even Captain Moore, whose expeditions brought the so-called "Great Pacific Garbage Patch" to public attention, seems swayed by the aesthetics of this enemy substance: I couldn't help but notice at the event that he was wearing a necklace made out of pieces of plastic. This experience sparked me to think of how other artists such as Mark Dion and Stuart Haygarth similarly cluster salvaged plastics by type, size or shape, by colour, and/or into colour patterns such as a rainbow: through collection and arrangement, they make beautiful what would otherwise be ugly. I will briefly describe exemplary work from each artist before discussing further.

For *Tate Thames Dig* (1999), which I had the chance to see at the titular museum in London in 2002 (just before entering academia), Dion

> spent two weeks with a team of invited archaeological experts and volunteers beachcombing on two sites on either bank of the river Thames. […] Objects were cleaned, catalogued and ordered according to type, weight and colour, and arranged within a large, doubled-sided cabinet of curiosities – a container with strong allusions to Renaissance and Victorian traditions of excavation, collection and display.
>
> (Viney 2010)

While plastic was not the only material collected, it was rather abundant given its lack of biodegradability. And this material is particularly prevalent in one familiar image of the piece: a drawer containing colourful buttons, toothbrushes and swizzle sticks, among other objects. Dion has carried these efforts forward: in 2013 he joined the Expedition Gyre, a group of biologists, educators and artists investigating the Great Pacific Garbage Patch. A photo by Kip Evans published in *DIS* magazine (Lookofsky n.d.) shows the artist kneeling on a black sand beach with an array of debris – bucket lids, laundry detergent caps and sports drink lids – arranged before him by colour and type.

Stuart Haygarth's *Tide Mark* (2004) is part of a larger project called *Dungeness*, after the Kent, UK-area headlands beach from which the artist collects materials. Mounted on a wall behind glass, *Tide Mark* consists of plastic debris organized by colour – including cigarette lighters, razors, toothbrushes and a surprisingly large amount of jelly shoes. As Haygarth describes it in an artist's statement, "*Tide Mark* is a collection of primarily plastic objects categorized by colour. Starting with white objects and ending with black, a kind of tide mark through the colour spectrum is produced" (n.d.[a]). Like Dion, Haygarth has continued work in this vein, as with 2011's *Tide*, an update of a 2005 piece. The object in question is spherical and functions as a chandelier; his statement explains that "this piece is created totally from clear colourless plastic collected from the beaches of the United Kingdom. The plastic objects are cosmetically transformed into sandblasted glass elements produced by the abrasive nature of the sea water and sand" (n.d.[b]).

As much has been said about both *Thames Tate Dig* and *Tide Mark*, I will summarize some basic points. First, both works have a sort of ironic thread running through them: they treat plastic and disposable objects reverently, as if they were valuable artefacts – which reminds us of the crucial irony that these items are not, in fact, waste if they can never really be thrown away – while also modelling a form of care over consumption. Second and relatedly, both artists invoke a systems perspective: they make clear the closed-system reality of our planet, in which the unseen is never actually gone. Finally, both artists engage in a kind of archaeological gleaning process[4] that indirectly highlights the concept of the Plasticene. Indeed, their work seems to anticipate science reporter Christina Reed's imagined scenario:

> One million years from now, geologists exploring our planet's concrete-coated crust will uncover strange signs of civilisations past. 'Look at this,' one will exclaim, cracking open a rock to reveal a thin black disc covered in tiny ridges. 'It's a fossil from the Plasticene age'.
>
> (Reed 2015)

But as suggested above, I'm most interested in how these artists make "sightly" – in both the sense of "visible" and of "beautiful" – what scientists have called the "unsightly problem" of plastic pollution (Law and

Thompson 2014, 144). From a straightforward activist perspective, there are some problems with this dynamic. As Melissa Potter notes in her review of "Consumer Culture," a 2012 exhibition at Chicago's Woman Made Gallery that featured the work of Ifeoma U. Anyaeji, discussed further below, "[o]verindulgence in aesthetic beauty runs the risk of sugarcoating the social and cultural concerns surrounding consumerism" (2012, 28). Further, plastic art might have the opposite of its ostensible intended effect of raising awareness about and, thus, reducing, plastic pollution; as environmental studies PhD student Patrick Chandler recounted to me during a video chat on 23 July 2019 of his time working for a plastic art foundation, audiences would often make jokes along the lines of, "I guess it's okay that we have so much plastic, since we can just make art out of it!"

If the ostensible point is to draw attention to plastic pollution, why make it look appealing instead of terrible? If Dion's job is, as he puts it, to "chronicle disaster" (Lookofsky n.d.), why is his work so pleasantly ordered, organized and controlled? But perhaps that is *not* the point – perhaps the point is to articulate our emotional and material embeddedness in plastic. By this view, these artists are not hypocrites, and they are not romanticizing problematic objects, exactly. Rather, they demonstrate larger truths: the attraction to plastic, and the impulse to aestheticize it, is part of human nature. Or, perhaps, animal nature: we find pleasure in collection, curation and colour, not unlike many species of birds, such as the bowerbird who makes his nest with colourful materials, often plastic.[5] And just as birds like the Midway albatrosses have trouble distinguishing between what is food and what is trash – thus accidentally ingesting plastics – we blur the lines between trash and art by turning the former into the latter. I thus identify in these artists' work a deeply ambivalent scenario: aestheticizing plastic is a way of living both against and with it.

Boetzkes' work proves crucial here again, outlining the contradictory and ironic character of plastic. She theorizes plastic as "an unprecedented and particular form of waste that is not waste at all [...] The original informational richness of plastic that destines it for disposal reverses so that plastic becomes its opposite; it becomes recalcitrant, toxic, unwieldy" (2019, 185). She concludes that "[t]he ubiquity of plastic [in art] is evidence of the state of paralysis in the face of the ecological condition. It airs a negative dialectic: what art reveals is precisely the inefficacy of acts of revealing" (184). While I agree with her line of thought more broadly, I am suggesting that revealing is not the sole or even the primary purpose of such art. Indeed, I have argued elsewhere that a focus on didacticism – an artwork's ability to teach its audience about environmental problems – has obscured the many other potentials for environmental art beyond revelation or education. These potentials include

> expressing dissatisfaction or disaffectation with the environmentalist status quo, bearing witness to crisis, enacting catharsis, raising activist morale, building community, serving as cultural diagnoses, indexing

> and helping us understand our current eco-political moment, "separat[ing] [ourselves] from [...] obligatory response[s]" to disaster (Kuipers 2005, 21), mitigating the "partisan divide over environmental matters that has [...] hardened [over the past fifty years] into a [...] wall of bitterness" (Souder 2013, 335), inculcating a new range of responses to crisis, modeling flexibility and creativity in the face of crisis, and inspiring what geographer Shiloh Krupar calls "artful endurance" (2013, 25).
>
> (Seymour 2018, 7)

Perhaps, then, the point of such art is simply to embody tension, to embody ambivalence.

Feminist Art of the Plasticene: Peché and Anyaeji

What is arguably missing in the work of Jordan, Dion and Haygarth is a sense of social and environmental justice; their found-object style can inadvertently absent the human relationship to plastic and its differential effects and thereby produce a kind of flattened-out, universalist perspective. Further, their archaeological/expedition process may, for some, echo problematically masculinist approaches to the environment.[6] In what follows, I will explore how Ruth Peché and Ifeoma U. Anyaeji differ – physically transforming plastic and, in the process, reflecting on plastic's gendered, patriarchal and colonial properties.

Like Jordan, Dion, Haygarth and others, Peché's work often entails the organization of plastic materials by colour. Indeed, one of her major works, installed at the University of Alcalá in July 2018 as part of the International Conference on Environmental Humanities, is titled *rainbow*, and its bright interlinked pieces are arranged accordingly (Figure 4.1). When I was lucky enough to participate in a workshop titled "Plastic Botany/Botánica Plástica" with Peché as part of the conference, it began, not surprisingly, with us sorting materials by colour. But as suggested above, Peché departs from the likes of Jordan, Dion and Haygarth in how she transforms plastic materials – and in how she sees herself as "intercepting [plastic on] its way to the waste stream" ("Statement" n.d.) – rather than excavating, "discovering" or reclaiming it after the fact.[7] As she explained her process to me in an email interview from 23 September 2019,

> First I collect, clean and classify the plastic bags, by thickness and color. Then I cut them in spiral[s] to make [...] plastic yarn so I can weave and crochet with it. I crochet [...] intuitively in whatever shape I feel that time and then I sew them together. The only patterns I follow are nature shape patterns.
>
> (Peché 2019)

Figure 4.1 Ruth Peché installing *rainbow*, 2018.

For example, *rainbow's* interlinked crocheted pieces mimic corals and sea anemones (Figure 4.2).

In our interview, Peché described how these qualities of her work might be considered ironic – "As an artist I am urged to transform [plastic] into beautiful organic and natural shapes that are endangered by the material itself" – as well as shot through with ambivalence – "[I am] working with love with a material I hate". In this way, Peché represents Boetzkes' idea that plastic is "much more than a mere substance, pollutant, or metaphor of postmodern superficiality. Rather it is a pervasive condition that produces conflicted relationships, behaviors, and affective modes" (Peché 2019, 182).

Figure 4.2 Close-up of Ruch Peché's *rainbow*, 2018.

But perhaps more positively, we can see how "working with love with a material [you] hate" models a different approach, if not a solution per se, to plastic. In fact, Peché has devised a manifesto that consists of three conceptual moves: "*Transformar* (Transform)/*Dotar de valor* (Endow with value)/*Ya no es basura* (It's not trash anymore)" ("Mutaciones Plásticas" 2018).

Importantly, these ways of giving value to waste also function to give value to feminized aesthetic practices. As Peché told me in our 2019 interview, reflecting on her plastic crochet technique, "weaving, sewing, embroidery or crocheting have been tasks traditionally linked to [women]"; for many, these activities were "the only means of creative expression they were allowed to engage [in]". In a separate presentation for the aforementioned conference, Peché also mentioned that she sometimes begins by fusing bags together using an iron ("Mutaciones Plásticas" 2018) – redolent of women's traditional laundry duties. But more specifically, regarding the material realities of plastic, she stated in our interview:

> [Women] are given (imposed in most cases) the responsibility of taking care of others and our home, therefore our environment. And I feel that the use of plastic is imposed [on] us in our daily life by the big food companies and huge oil corporations. [...] [These] big corporations that dictate our way of life are in the hands of powerful men. So in the end [plastic] is a question of masculine and money power.
>
> (Peché 2019)

These points echo those of Patrick Chandler, who noted during our 2019 video chat that the story of plastic exemplifies the patriarchal-capitalist mindset in so far as it entails aggressively maximizing profit with little attention to consequences.

Ifeoma U. Anyaeji's colour palette is relatively muted compared to Peché and the other artists I have discussed; she largely eschews the colour groupings I have described above and employs more whites and pastels. *Cumulus* (2013), for example, consists of long braids of plastic looped into individual circles and then enclosed with longer braids in order to form, as the title implies, a cloud-like structure. The dominant colours of this piece are white, grey, dull yellow and dark brown. But like Peché, Anyaeji wants to "exten[d] the significance for objects assumed to have lost their 'newness'" ("Artist Statement" n.d.), including her main media of plastic bags and bottles. And like Peché, Anyaeji draws on typically feminized aesthetic practices to do so. As she explains her unique process in her artist's statement:

> Using a concept I developed, called Plasto-art [...] I repurpose these bags and bottles with a receding traditional Nigerian hair-plaiting technique, called Threading (also know[n] as Ikpa Owu – in Igbo language). I visually express the narrative of this domestic object's possible transition from discarded to the aesthetic or functional – an evolution from redundancy to utility while retaining the hair craft's technique or object's physical state.
>
> ("Anayaeji" n.d.)

The artist thus, ironically, extends time and care to a problematic material rather than rejecting or demonizing it. And she does so specifically through a devalued technique related to Black women's hair – itself a site of devaluation and stigmatization, especially in the United States.[8] We might therefore understand Anyaeji's work within the larger context of "Black feminist eco-art" (Frazier 2018, 173), and see both Anyaeji and Peché as modelling responses to the question posed by rhetoric scholar Christine Harold: "how might we manage the material excesses of consumer capitalism by building on, rather than repudiating, our attraction and attachment to objects" (2020, 18) – and to plastic in particular, with its incredible flexibility, tenacity and capacity for colour?

Finally, I see Anyaeji's work as responding to the issue of "waste colonialism," in which the (supposed) disposal of materials such as plastic relies on the existence of colonized land, in which developing nations are deluged by waste exports from developed nations, and in which waste disproportionately affects marginalized populations (Liboiron n.d.). The artist observes that her chosen media of plastic bags and bottles are "two main global environmental pollutants, especially in [her] home country Nigeria" ("Artist Statement" n.d.). Without a strong recycling infrastructure for such materials, as reporter Afeez Hanafi explains, seasonal rains in cities such as the capital of Lagos can lead to disastrous floods, as "plastic bottles, plastic bags [...] and other waste materials cra[m] drainage channels, thereby hindering free flow of water" (2018). Apart from death and property damage, such floods can bring "water-borne diseases such as cholera, typhoid and dysentery" (Hanafi 2018). We can therefore understand Anyaeji's work as a commentary, albeit very subtle, on "the violent White, colonial cis-heterocapitalist logics that continue to underpin ecological [degradation]" (Frazier 2018, 170), and on the differential impacts of plastic in particular.

Conclusion

I want to locate the theorizing and artworks I have discussed above in the larger context of what I have recently termed "bad environmentalism": the embrace by contemporary environmental artists and activists, especially queer ones, of complex affects – not just ambivalence, but also irony, awkwardness, absurdity and perversity. This embrace opposes the strident moral conviction that we've historically seen within Western, largely white, mainstream environmentalism, as well as its cloying sentimentality and sanctimony. This kind of embrace is not, as I hope to have suggested above, unthinking or uncritical. I turn back to Katie Schaag here: in an essay on her university course centred on plastic, she reflects, "[r]ather than attempting to resolve plastic's conflicting associations or only focusing on plastic's toxicity, I designed the course to instigate critical reflection on the complex implications of plasticized cultural, environmental, and aesthetic forms" (2018). Meanwhile, Chelsea M. Frazier asks, more pointedly, "how might we continue to mine Black feminist art, not for resolutions to poorly conceptualized problems, but for better understandings of our hopeful, yet damaged presents?" (2018, 187). Plastic, whether in the figurative hands of theorists or the literal hands of artists, both engenders and encodes uncertainty and irresolution. It speaks to the deep ironies and ambivalences of our current moment.

But to go further, perhaps what we are also seeing here is the insidious character of plastic materiality and marketing; how plastic *capitalizes* on the aforementioned ambivalence. It makes itself feel inevitable, in a way that Trisia Farrelly's matter-of-fact statement – "glitter [is] microplastic" (Gabbatiss 2017) – ironically echoes. As Amanda Boetzkes mused to me in an email on 3 March 2020,

> Perhaps we only have the thought of loving [products such as glitter] when we feel we have to give [them] up for environmental reasons. [...] How would we feel about [them] if we felt that plastic was easily replaced with something else?
>
> (Boetzkes 2020)

Would our love feel so urgent, or our hate so passionate? If plastic exploits our ambivalence in order to continue its astonishing spread, then it is those mixed feelings that most require our attention now.

Acknowledgements

Conversations with Vanbasten Noronha de Araújo and Patrick Chandler, and generous draft feedback from Amanda Boetzkes, were invaluable in helping me develop this essay. I am also grateful to Ruth Peché for sharing her practice with me and others in her workshop (including Erika Bsumek and Katharina Kalinowski), and for conversing with me after. Finally, thank you to the organizers of the International Conference on Environmental Humanities at the University of Alcalá and to the Rachel Carson Center for Environment and Society for publishing my early thoughts on this topic on their *Seeing the Woods* blog.

Notes

1 As Vanbasten Noronha de Araújo observes, drawing on a 2019 BBC report (https://www.bbc.com/news/av/uk-england-manchester-47789342/biodegradable-glitter-backed-by-sir-david-attenborough), glitter constitutes less than 0.1 per cent of the plastic material in oceans (2019, 26).
2 See https://discardstudies.com/2014/05/07/why-discard-studies/.
3 Haram et al. trace the use of the term "Plasticene" to palaeoclimatologist Curtis Stager in 2011, who was riffing on blogger Matt Dowling (2020, 2).
4 As Viney sees *Tate Thames Dig*, "it is a piece that collects, sorts and displays objects to enact both the history of museums and to criticise how this history has taken shape" (2010).
5 I first presented these ideas in a blog post, "The Aesthetics and Affects of Art in the Plasticene". See https://seeingthewoods.org/2019/05/07/uses-of-environmental-humanities-nicole-seymour/. See also https://animals.sandiegozoo.org/animals/bowerbird.
6 Art critic Iwona Blazwick describes how Dion has "taken on the role of explorer, detective, natural scientist, and museum curator" (2001, 103). While Dion, importantly, relied on amateur volunteers from local community groups to collect the items for *Tate Thames Dig*, this process is not referenced in the artwork itself.
7 Peché is part of a collective of artists, designers and architects known as Project Vortex; its members "actively intercept[t] the plastic waste stream as part of [their] independent practices" ("Who We Are" n.d.). See https://www.projectvortex.org/the-project-vortex-collective.
8 See, for example, Ingrid Banks' *Hair Matters: Beauty, Power, and Black Women's Consciousness* (2010) and Althea Prince's *The Politics of Black Women's Hair* (2010).

References

Ah-King, Malin and Eva Hayward. 2013. "Toxic Sexes: Perverting Pollution and Queering Hormone Disruption." *O-Zone: A Journal of Object-Oriented Studies* 1: 1–12.

Anyaeji, Ifeoma U. 2013. *Cumulus*.

Anyaeji, Ifeoma U. n.d. "Artist Statement." https://www.ifeomaanyaeji.com/.

Azzarello, Robert. 2012. *Queer Environmentality: Ecology, Evolution, and Sexuality in American Literature*. New York: Routledge.

Banks, Ingrid. 2010. *Hair Matters: Beauty, Power, and Black Women's Consciousness.* New York: New York University Press.

Blazwick, Iwona. 2001. "Tate Thames Dig." *Oxford Art Journal* 24 (2): 105–112.

Boetzkes, Amanda. 2019. *Plastic Capitalism: Contemporary Art and the Drive to Waste.* Cambridge, MA: The MIT Press.

Chen, Mel. 2012. *Animacies: Biopolitics, Racial Mattering, and Queer Affect.* Durham, NC: Duke University Press.

Davis, Heather. 2016. "Imperceptibility and Accumulation: Political Strategies of Plastic." *Camera Obscura* 31 (2): 187–193.

Dean, Tim. 2006. "The Antisocial Homosexual." *PMLA* 121: 826–828.

de Araújo, Vanbasten Noronha. 2019. "Shimmery Waste: A Queer Critique of the Narratives on Glitter Pollution." Master's thesis, Central European University.

Dion, Mark. 1999. *Tate Thames Dig.* London: Tate Modern.

Frazier, Chelsea M. 2018. "Thinking Red, Wounds, and Fungi in Wangechi Mutu's EcoArt." In *Ecologies, Agents, Terrains*, edited by Christopher P. Heuer and Rebecca Zorach, 167–194. New Haven, CT: Yale University Press.

Freinkel, Susan. 2011. *Plastic: A Toxic Love Story.* Boston, MA: Houghton Mifflin Harcourt.

Gabbatiss, Josh. 2017. "Glitter Should Be Banned over Environmental Impact, Scientists Warn." *The Independent*, November 16. https://www.independent.co.uk/environment/glitter-ban-environment-microbead-impact-microplastics-scientists-warning-deep-ocean-a8056196.html.

Hanafi, Afeez. 2018. "Plastic Pollution: Nigeria's Untapped 'Waste Wealth' Fuels Environmental Disaster." *Punch*, August 11. https://punchng.com/plastic-pollution-nigerias-untapped-waste-wealth-fuels-environmental-disaster/.

Haram, Linsey E., James T. Carlton, Gregory M. Ruiz, and Nikolai A. Maximenko. 2020. "A Plasticene Lexicon." *Marine Pollution Bulletin* 150: 1–4. https://www.sciencedirect.com/science/article/pii/S0025326X19308707.

Harold, Christine. 2020. *Things Worth Keeping: The Value of Attachment in a Disposable World.* Minneapolis: University of Minnesota Press.

Haygarth, Stuart. 2004. *Tide Mark.* http://www.stuarthaygarth.com/works#/tidemark2004.

Haygarth, Stuart. n.d.(a). Untitled artist's statement about *Tide Mark.* http://www.stuarthaygarth.com/works#/tidemark2004.

Haygarth, Stuart. n.d.(b). Untitled artist's statement about *Tide.* http://www.stuarthaygarth.com/works#/tide2011.

Jordan, Chris. 2010. *Albatross Mandala Midway Island.* https://jamestowngazette.com/rtpi-presents-albatross-film-plastic-pollution-forum-april-23/.

Jordan, Chris. 2009. *Message from the Gyre* series. http://www.chrisjordan.com/gallery/midway/#CF000313%2018x24.

Jordan, Chris, dir. 2017. *Albatross.*

Law, Kara Lavender and Richard C. Thompson. 2014. "Microplastics in the Seas." *Science* 345 (6193): 144–145.

Liboiron, Max. 2013. "Plasticizers: A Twenty-First Century Miasma." *Accumulation: The Material Politics of Plastic*, edited by Jennifer Gabrys, Gay Hawkins, and Mike Michael, 134–149. New York: Routledge.

Liboiron, Max. n.d. "Waste Colonialism." *Discard Studies.* https://discardstudies.com/2018/11/01/waste-colonialism/.

Liboiron, Max, Manuel Tironi, and Nerea Calvillo. 2018. "Toxic Politics: Acting in a Permanently Polluted World." *Social Studies of Science* 48 (3): 331–349.

Lookofsky, Sarah. n.d. "Mark Dion: Trash on the Beach." *DIS*. http://dismagazine.com/disillusioned/52015/mark-dion-trash-on-the-beach/.

Meikle, Jeffrey. 1995. *American Plastic: A Cultural History*. New Brunswick, NJ: Rutgers University Press.

Peché, Ruth. 2018. "Mutaciones Plásticas." Paper presented at the International Conference on Environmental Humanities, July 4. University of Alcalá, Spain.

Peché, Ruth. n.d. "Statement." http://www.ruthpeche.com/ruthpeche/statement.html.

"Plastic Pollution." n.d. *Greenpeace.org*. https://www.greenpeace.org.uk/challenges/plasticpollution/.

Potter, Melissa. 2012. "A Consuming Conundrum." *Afterimage: The Journal of Media Arts and Cultural Criticism* 40 (2): 28–29.

Prince, Althea. 2010. *The Politics of Black Women's Hair*. Toronto: Insomniac Press.

Project Vortex. n.d. "Who We Are." https://www.projectvortex.org/the-project-vortex-collective.

Reed, Christina. 2015. "Plastic Age: How It's Reshaping Rocks, Ocean and Life." *New Scientist*, January 28. https://www.newscientist.com/article/mg22530060-200-plastic-age-how-its-reshaping-rocks-oceans-and-life/.

Schaag, Katie. 2018. "The Pleasures of Teaching Plastic." *Edge Effects*, October 2. https://edgeeffects.net/teaching-plastic/.

Schaffer, Guy. n.d. "Camp." In *Discard Studies Compendium*, edited by Max Liboiron, Michele Acuto and Robin Nagle. https://discardstudies.com/discard-studies-compendium/#Camp.

Sheridan, Kate. 2017. "Glitter Is an Environmental Scourge That Is Wrecking the Oceans. Should It Be Banned?" *Newsweek*, November 29. https://www.newsweek.com/glitter-environmental-scourge-wrecking-oceans-should-it-be-banned-725826.

Seymour, Nicole. 2018. *Bad Environmentalism: Irony and Irreverence in the Ecological Age*. Minneapolis: University of Minnesota Press.

Seymour, Nicole. 2020. "Queer Ecologies and Queer Environmentalisms." In *The Cambridge Companion to Queer Studies*, edited by Siobhan Somerville, 108–122. Cambridge, UK: Cambridge University Press.

Viney, William. 2010. "Mark Dion and 'Tate Thames Dig' (1999)—An Extract." *Waste Effects*, May 24. https://narratingwaste.wordpress.com/2010/05/04/mark-dion-and-tate-thames-dig-1999-an-extract/.

Whitehouse, Penny. n.d. "DIY Biodegradable Glitter for Kids." *Mother Natured*. https://mothernatured.com/nature-crafts/diy-biodegradable-glitter/.

Willingham, A.J. 2017. "Glitter is Not Just Annoying, It Could Be Bad for the Environment." *CNN*, November 29. https://www.cnn.com/2017/11/29/health/glitter-environment-hazard-microbead-ban-trnd/index.html.

5 Political Drugs

Materiality in *Testo Junkie*

Kathrin Peters

The Pleasure of the Text

A text is an object. It can be a fetish object, as Roland Barthes writes. As a reader, I do not have this object at my disposal; rather, it entangles me with its chicaneries: "vocabulary, references, readability, etc." (Barthes 1975, 27). The text fills the reader with fantasies about the author – a figure, Barthes continues, who is "lost in the midst of a text" but is neither a representation nor a projection. That's what this essay is about: the vocabulary, references, readability, etc., of the book *Testo Junkie* and the figure who is constructed in the book as "I," but who neither represents Paul Beatriz Preciado as a person nor exists as sheer projection of myself as a reader. This figure emerges in the process of writing and reading, into which numerous other texts and images are included. One could also say, in anticipation of my thoughts below: "the author is a materialization". Yet Preciado's text itself is a material object, a theory book that is *dirty* in every respect, even in its form. *Testo Junkie: Sex, Drugs, and Biopolitics in the Pharmacopornographic Era* (2013) is the protocol of a self-experiment with testosterone, a discourse history of hormones, and a postporno. It has prompted me to once again ask about the relationship between materiality and gender. This is a long-standing question in gender studies that, in view of today's diagnosis of "pharmacopornography," I am posing differently, because "materiality" is understood here to encompass not only bodies but also substances that penetrate them. What's the effect, what is at work here?

To this, I also add books from other shelves: the *Vernon Subutex* novels (Despentes 2015–2017). The author of these books, Virginie Despentes, has designed a heap of figures so fitting that I repeatedly thought I know this or that person, or even that I *am* that person. Probably because all the current fuss about identity and identification is a constant theme of the trilogy. That said, it's a savvy choice for the story to be mainly about *dis*identification and *de*subjectification – apart from the plot about the record store owner Vernon Subutex, who has become homeless in gentrified Paris, and an entourage of people who at first want nothing to do with him, then urgently want something from him. Here, already, the book is about sex, drugs and

DOI: 10.4324/9781003023319-8

biopolitics; Vernon Subutex is the name of a heroin substitute. I admire Virginie Despentes for her astuteness in placing this male seducer figure, desolate and delirious as it is, at the centre of attention, and in arranging all sorts of female and queer figures around him, or at least supposedly doing so. The strategy worked: the books have become bestsellers. From the perspective of the Despentes universe, Preciado's *Testo Junkie* reads slightly differently, because *Testo Junkie* is also about Virginie Despentes (here, she is called VD), with whom Preciado had a relationship as he was writing the book. A true pleasure of this text overcame me.

Pharmocopornographic Regime

Testo Junkie is therefore many texts – a "testosterone-based, voluntary intoxication protocol," a "body-essay," a "somato-political fiction" or a "theory of the self, or self-theory" (Preciado 2013, 11). A theory of the self in the wake of Michel Foucault, but also a self-theory that, as I understand it, concerns the self or the self-understanding of theory. This shaking of the self of theory arises, above all, from the textualization of experiences on testosterone, which Preciado is using to upgrade, prosthetically, his/her "low-tech transgender identity". But the seriousness of this theory is questioned not only because the observer and the observed coincide in these passages, [1] but also because they evoke the question: under the influence of which drugs and hormones are theory texts otherwise written?

However, a "gender reassignment" is not the goal of these testosterone treatments, because that would be a goal within a system of gender normativity, committed to the idea of a true gender that Preciado does not have in mind. Self-treatment with testosterone is not about a change to the other side of the binary order of gender but about expanding, for testing purposes, the register of materials that designate bodies as sexual: from a body that shapes itself through hairstyles and shaving, clothing and perhaps body treatments, that is through *low-tech* means, to a chemically altered body, one that is perhaps not *high-tech* but nevertheless molecularly altered.

> I put a fifty-milligram dose of Testogel on my skin, so that I can begin to write this book. It isn't the first time. This is my usual dose. The carbon chains [...] gradually penetrate my epidermis and travel through the deep layers of my skin until they reach the blood vessels, nerve endings, glands. I'm not taking testosterone to change myself into a man or as a physical strategy of transsexualism; I take it to foil what society wanted to make of me, so that I can write, fuck, feel a form of pleasure that is postpornographic, add a molecular prosthesis to my low-tech transgender identity composed of dildos, texts and moving images [...]
>
> (Preciado 2013, 16)

The text may play with the genre of diary-like self-assurance, a genre that has constituted subjectivation as a transsexual since the early twentieth century (Runte 1996). But at the same time, this is precisely what the text is *not*, because *Testo Junkie* does not operate an identity discourse that is able to play an important, even existential role for the ambivalent and painful subjectivation of trans persons; in Preciado, identity is explicitly understood as *composed* – of things, texts and images. These include the package inserts for the testosterone preparation, which proclaim that the drug is reserved for the use of adult males; the dildos that are repeatedly mentioned; the drag king-workshops that Preciado attends;[2] pornography and, even more so, postpornography not oriented towards genitality, towards penetration and a phallic order. What is observed on one's own body is individual *and* a political project, because "the change happening in me is the metamorphosis of an era" (2013, 22).

Preciado has given the name of "pharmacopornographic regime" to what distinguishes our present time. He thereby overwrites Deleuze's concept of the "control society," since "control society," he argues, did not take into account the connection between body, sexuality and technology – the gradual incorporation of technology; it would, to put things pointedly, be thought too immaterially: "[I] shall call it the pharmacopornographic society. A politically programmed ejaculation is the currency of this molecular-informatic control" (2013, 77).

In *Pornotopia*, a book about the magazine *Playboy* and its medial and architectural excrescences, Preciado (2014) traces how the currency of ejaculation made a start in the 1950s: a rabbit replaces the stag that the magazine, founded in 1953 by Hugh Heffner, initially wanted to use as a logo. But the stag turned out to be too paternalistic a figure, too much head authority with antlers and majestic gait. The aim was something else, a less paternal form of masculinity, more about a bunny among bunnies. But bunnies aren't the only thing to be seen in *Playboy*. Across many double-page spreads, spacious apartments and lofts are portrayed in detail, decorated with modernist furnishings, with Eames and Womb chairs. For the urban bachelor who had escaped from the family houses of the suburbs, this furniture served as a stage for virile masculinity. For the activities of this bachelor, modern architecture and interior design are far more than just a backdrop: with their modularity and mobility, the mechanisms for adjustment and rotation that distinguish modern design from the quite immobile furniture of bourgeois interiors, the swivel chairs, tilt sofas, fold-out kitchens, and rotating beds, are allegories of the manoeuvrability of the single male and, at the same time, serve as his centre of operations. A flow, choreographed by human and nonhuman actors, drives the *girls* into the apartment and, above all, back out again. Is that really how people lived? I doubt it – the costs seem too high to me, financially, but above all, emotionally. Spending and overspending seem to lead only to ruin. But what matters most is not the degree to which this image was realized. What is at stake is the emergence of a sexual-gender imagination that accompanies a specific masculinity and will have produced its reality effects, among other things, precisely in the failing of a continuous performance that was and is expected by all participants.

The spatial designs that drive this imagination, and are thus always designs of a subject, are meant above all to counter the danger of "emasculation," which might consist, for example, of regressing to a couch potato in front of the television, snacks always within reach, as Lynn Spigel (1992) concluded in reading discourses of the postwar period. The new figure of the unattached and potent young man of any age is to be understood as distinct from the flaccid old patriarch (who nevertheless remains a constant danger of what the young man might become). But the true target of this figure's aversion is being gay. However, it is not always easy to draw this distinction discursively, because the open promiscuity around which *Playboy* ultimately revolves must be spelled out in heteronormative terms. Conversely, and today, it is easy to observe how, with some effort, any impression of promiscuity must be avoided in order to normalize being gay. Despite all these attempts at normalization, Preciado's point is that in the 1950s and 1960s notions of masculinity changed. This becomes clear not only in imaginations of space and design, but also in other *things*. Preciado's *Contrasexual Manifesto* forcefully formulates the disruptive power of the dildo "because it shows that both masculinity and femininity are subject to the social and political technologies of construction and control" (2018a, 59). The dildo – mutable, able to be bought and sold – is better than the phallus, which of course is always only symbolic. And in deconstructive parlance: the dildo is presented as the supplement that precedes the penis, denaturalizing it.

The pharmacopornographic subject is governed by a constant "chain of excitation-frustration". The permanent excitement, stimulated by images, texts, fabrics, bodies and digital technologies, and aiming to achieve constant performance defined by achievement, duration and overpowering force, is regularly followed by collapse. Burn-out, impotence, or over-indebtedness are the modes of collapse; they occur psychically, physically and financially. Numerous narratives and technologies have emerged around these various forms of collapse, as well, offering prevention, self-optimization, or course of treatment, but they ultimately only serve to keep the chain reactions going. Being weary is part of the pharmacopornographic regime and cannot be a lever of criticism for Preciado.

This chain reaction of excitement and frustration that thus began in the 1950s is driven by pharmaceuticals that penetrate the body in one way or another: synthetic hormones, Viagra or contraceptives. Yet antidepressants, Ritalin, or speed also keep the excitation-frustration chain in motion. There are actually no more bodies left that are not drugged in one way or another, even if only involuntarily due to residues in drinking water and food. Naming antidepressants and contraceptives in one breath suggests that, despite all the differences in their fields of application, both are aimed at the production of invulnerability and independence (as I would call it). Without in any way romanticizing a life with manifest depression or unwanted pregnancies, the point here is to focus on the intertwining of these medications.

Preciado's analysis is sober: these biochemical technologies are productive in Foucault's sense. For Preciado, the subjects have no direct *agency*; rather, it is the substances "that feed the subject's actions and through which they turn into agents" (2013, 35). Artificial testosterone belongs firmly in the arsenal of the pharmacopornographic regime. It is used for athletes, for men with erectile dysfunction, in body building or for gender reassignment, more or less legally, depending on the case. Its use by women is forbidden, since oestrogen is reserved for them: first, for contraception, and later during menopause. The reproductive cycle of the female body is regulated by oestrogen medications to such an extent that one must speak of a "construction of biofeminity" (Preciado 2013, 212).[3]

Preciado gets testo-gel through friends, acting as a gender-hacker within a transgender black market: "it is a bit like biting the dick that's raping you" (2013, 61). After s/he applies it, the sweat takes on a sweet smell, s/he is unable to sleep, a compulsion takes over the ego of the text and the level on which we find ourselves remains not entirely clear: whether as fiction or a report, assuming the two can be precisely distinguished.

Materialization and Performativity

The question of whether the gendered body is – in a forced set of alternatives and simplified argument – natural *or* constructed is of course one that has been debated tirelessly, in gender studies and many other fields; it is in particular in the movements that declare themselves to be "against gender" that the worst polemics concerning the constructedness thesis are to be found. Problematizing the body as the origin and foundation of gender broke with a common sense understanding from the postwar period: the distinction between sex as something biological and gender as something social, together with the rhetoric of nature and culture that was embedded in this discourse. The sex/gender system as model that women's studies were built upon until the late 1980s, when Judith Butler and Teresa de Lauretis began to critique it, is "not a concept, it is an ideology" for Preciado, "*it is a techno-political ecology*" (2008, 112, emphasis in original; Preciado 2013, 107).[4] The crucial point is that the sex/gender system was first developed in medical treatment for intersexuality and transsexualism beginning in the 1950s by John Money. The term "gender" was coined to define socially and culturally recognized groups of "masculine" and "feminine" behaviour (106). The idea of a physical materiality of sex on which social and cultural constructions of gender are based thus arises against the background of a surgical and hormonal alteration of the gendered body.[5] The medical-psychiatric speculation goes something like this: since the self-understanding of a gendered being develops from bodies, operations and hormone treatments entail a corresponding gendered subjectivation. After sex and gender had been separated conceptually, both can be brought into normalizing consensus by hormone pills and genital surgery. Gender-as-construction is often a violent approach of sexology and sexual medicine to adapt the body to a "true sex" and vice versa. To call the sex/gender system an ecology points to the interdependence of technology, gender politics, somato-physical materiality and cultural imagination.

For Preciado this is reason enough to understand sex as constitutively technological – in an entirely operative sense. Seen thus, the drag king practice that Preciado pursues in workshops as a participant and leader extends into the body, as "biodrag" (2013, 191). The material intervention in the body is a performative-subversive strategy, a gender hacking: substances are used contrary to their medical indications. The medical promises of plastic surgery and hormone treatment to shape sexual characteristics are taken at face value, as it were, but the strictly regulated provision of hormones, which stipulates that treatment may only take place within the framework of a binary sexual order, is subverted. The provisionality and instability of every "natural" order embedded in sexual medicine is emphasized and played out.

As plausible as I find this strategy of affirmative criticism to be, it also raises questions: these concern the concepts of materiality and performativity that Preciado references positively, but which, at the same time, seem to me to be somewhat imprecise. My point is not that a brilliant text like *Testo Junkie* ought to be measured against standards that miss what the text is after. The text is, of course, a performative text that does not aim to present a metatheory. However, Preciado is also a theorist who is too adept for it not to be worthwhile for us to follow the gender-theoretical work of his text more closely. These questions stay with me because Preciado's text cannot be recuperated into familiar narratives. *Testo Junkie* is not written from the perspective of an author struggling for truth and selfhood, although there are features of this, nor does it propagate the celebration of a "gender carnival" that would define gender as an act of choice and of a well-equipped wardrobe or suitable dildo. This has been a frequent misunderstanding in the reception of Judith Butler's texts on gender theory, which we find aspects of in Preciado, too. But *Testo Junkie* is too dark, ambivalent and existential to be suitable for a self-empowering reading. It is, not least of all, a letter to a deceased friend, a mournful text that begins the book. Still, isn't there something voluntary about biodrag? Is it possible that the book pays homage to a thinking of gender as substance, and not because it assumes a "naturally gendered body" but because the testosterone so obviously *works*. "Nothing but the feeling of strength reflecting the increased capacity of my muscles, my brain. My body is present to myself" (Preciado 2013, 21).

Butler's earlier theses, as well as aspects of more recent theories of materiality, can help here. It goes without saying that a subdivision into old vs. new, cultural constructivism, on the one hand, and material agency, on the other, cannot be maintained. Butler's intervention against what she calls an essentialism of gender or a metaphysics of substance draws from the experience of exclusion and marginalization within feminism, whose tendency to always know too precisely just what constitutes women and men runs the risk of confirming a heteronormative framework in which lesbians and gays, individuals who are intersexual or transsexual, do not fit. Butler develops the argument that the body is not suitable as a starting point for gender via her conception of materialization. Materiality and intelligibility cannot be separated, as Butler (1993) argues with recourse to Aristotle. Rather, matter is only ever recognizable and legible (intelligible) within cultural systems of meaning. Even the assertion of extradiscursivity is discursive.

In view of the many misunderstandings about gender, to which Butler's texts have themselves also certainly contributed, it is and remains important to emphasize that the issue here is not and cannot be denying materiality; Butler, too, knows clearly that bodies live and die. Butler's early texts (to which Preciado refers) may indeed omit questions of bodily experience and affectivity. But her decisive argument has not lost any of its persuasive power, for she is interested in understanding the body as pervaded by powerful, normative formations and not as a refuge for authenticity. As a thorough reader of Foucault, who in his discourse analysis of sexuality sees the body as pervaded by power, Butler considers the sexual body to be emphatically constituted by historical interpretations: "'Materiality' designates a certain effect of power, or, rather, *is* power in its formative or constitutive effects" (1993, 34). Bodies become socially visible and legible in a process of materialization, and in this respect, they *matter*. It appears to be impossible to hit upon sheer materiality prior to any marking – this is Butler's argument.

The concept allowing Butler to deconstruct "sex," or the gendered body, as the foundation of social reality, is that of performativity. If gender appears as a prior, naturally given fact that lies beyond every history or culture, this is precisely the result of a speech act: "signifying acts" that are constantly repeated and invoked. Butler's well-known example of gender performativity is drag performance. And yet for all the vividness of this example, its reception has very much brought along a number of constrictions. First, it has led to the misunderstanding that gender performativity is generally a deliberate act, as if one were to choose this dress one day and that dress the other. The existential subjectivity that is meant by being a sexual, gendered being, which is what Butler is theorizing, is not the same as cross-dressing. Second, the example of drag suggests that performativity goes hand in hand with performance, that is, that the implementation of gender significations is linked to a theatrical enactment, which, in order to have a subversive effect, ought to be as close to parody as possible.[6] Drag is, as Butler explains in retrospect, less a parody than what Butler recognizes here as an allegory demonstrating how norms are reproduced while also, in the same moment, being able to be contested (2004, 204–231). It's worth keeping this in mind: if we think we see a man dressed as a woman or a woman dressed as a man, then we assume that this person has a real and a false gender – that they are "in reality" a man or a woman. But what actually gives us, asks Butler, the certainty that there is a true or authentic gender? Is it the way one's clothes are worn, how one's hair is done? Is it because of the anatomy of the body? Are we making conclusions about genitals, although we don't even know what they might prove or whether they might prove anything? And why are we even assuming an either-or? For Butler, drag reveals gender norms alongside their ambiguity and unreliability. This is exactly how *Testo Junkie*'s concept of drag works: by observing the gender norms within which every subject necessarily acts while at the same time challenging them.

(New) Materialism

Testo Junkie is situated knowledge, and Donna Haraway is, not without reason, one of the most important theoretical sources for Preciado. The text advocates with full force for a knowledge production that knows it can only ever be carried out by specific bodies, that is aware that all knowledge is prosthetically composed. The prosthesis here is testosterone. The text itself, too, is a prosthesis, in the polymorphic form in which it presents itself to us – to read and know.

Broadly speaking, new materialism (which learnt a lot from Haraway) understands matter not as passive and inactive but as active and potent. But in order to avoid the risk of reversing the anthropological centring on humans as the only acting beings – which would mean understanding matter as an entity with a direct efficacy – it is important to think relationality. This means starting from a structure of actants, artefacts, fabrics, spaces and things. Matter only emerges in these structures, in other words; matter is not given – it does not precede the structure. Rather, matter is put into effect only within and by relations. That, in any case, is the crux of the concept of intra-action, as Karen Barad conceived it: matter is inextricably linked to its measuring devices (2007, 189–222). In order to work out this productive and processual dimension of materialization, Barad, too, has taken up Judith Butler's concept of performativity, while at the same time finding Butler's concept of materiality to be inadequate, because she believes it cannot sufficiently explicate material phenomena and ultimately assumes an anthropological primacy. Using the example of ultrasound diagnostics in gynaecology, she argues that a linguistic-discursive understanding of materialization is not sufficient for understanding the processes of materialization. Feminist science and technology studies has shown, she argues, that material processes and discursive processes always work together: "[M]aterial constraints, the material dimensions of agency, and the material dimensions of regulatory practices [...] make the gender interpellation of the fetus through ultrasound technology different from a situation in which 'girling' begins at birth" (2007, 194).

According to Barad, Butler still owes an answer to the question of how, then, social norms precisely influence materiality.[7] For Butler, this is first and foremost a question of psychological identification, which Barad finds wanting in view of a machinery in gynaecology that produces the foetus as a subject whose vital interests then take precedence over those of the pregnant woman.[8] Barad moreover points to an involuntary ambiguity of reproductive medicine, which she finds very similar to that of sexual medicine: by isolating egg and sperm cells and uniting them in test tubes and under microscopes, as shown on numerous digital images, reproductive medicine ultimately, and constantly, exhibits the "unnaturalness" of heterosexual reproduction (Barad 2007, 219).

The philosopher Butler certainly does not engage with concrete processes of materialization in the same way that a theorist of science does in examining laboratory situations. It seems to me that this is less an omission or exclusion, as Barad insinuates, than that Butler is simply not so interested in these aspects. Her initial concern was to intervene in a feminist discussion that viewed the body as a refuge of femininity, as something that could be a point of reference for identification – to intervene with the argument that wanting to understand matter apart from all contexts of meaning is neither possible nor productive. The series of terms *materia*, *mater* and *matrix*, which connects femininity with naturalness and reproduction, repeatedly overwrites any positively intended reference to the body (Butler 1993). Barad, by contrast, insists on understanding matter as agential, because matter also deploys effects within the structures in which it occurs – like testosterone in the body. The fact that testosterone is coded as a "male" hormone and that its allocation is biopolitically regulated is, from Barad's point of view, absolutely germane, because "[m]atter is substance in its intra-active becoming – not a thing, but a doing, a congealing of agency" (2007, 210). In this respect, I see Barad's objections more as meaningful additions and continuations that make it possible to comprehend materialization in complex, technical structures which always exist under discursive conditions, and which, in turn, provoke discourses.[9] For the concept of materiality that can be developed with Barad does in fact make it easier to understand how discursivity and materiality come together in the process of materialization. It is not only the substance that yields the effect. Rather, it is also the discourses that suggest, frame, and make it possible to name effects. To be sure, chemicals can make beards grow and voices change, but there are also placebo effects and medications that fail to work.

Nelly Oudshoorn has developed a very similar argumentation in her archaeology of hormones. She dates the appearance of hormones in the field of knowledge much earlier (and the "pharmacopornographic era" would probably need to be seen as having a longer prehistory). As early as the 1920s and 1930s, research began to focus on the strengthening and alteration of sex, based mainly on hormonal rather than anatomical treatments.[10] Oudshoorn (1994, 145) has shown that the invention of hormones abandons the anatomical model of the two sexes in favour of a model of secretions and thus moves from a descriptive to an experimental approach that makes sex measurable and quantifiable. Endocrinology has since developed techniques for diagnosis and intervention that have become common medical practice. What biomedicine diagnoses, advises and treats, what it knows, is based on material conditions – through devices, technologies and practices:

> Moreover, as biomedical discourses are the product of material conditions, they have fundamental material effects as well. The biomedical sciences have a material authority that is manifest in the form of diagnostic tools, screening tests, drugs and other regulatory devices [...] a reality that should not be neglected in feminist studies of science.
>
> (Oudshoorn 1994, 13)

So what happens when Preciado describes the effects of testosterone? Does the substance have a sexualizing, possibly masculinizing effect? Or is it rather that the text basically exhibits a masculine subjectivation? This seems to me to be the case: it is not always possible to distinguish where the text is dealing with substantial and material changes and where it is the imaginations as part of the medical-cultural discourse that are having an effect: postpornographic narratives have been included in *Testo Junkie* to the same extent as have descriptions of effects that are not only influenced by chemistry but also by reading the package leaflet to the medication. How might one also write outside these discourses? Testosterone works, has an effect, but not causally or deterministically. It works, too, as a text – as a text object that tells of fetishes and is itself one.

The "I" observes the hormone-induced somatic and psychological changes in its own body, as well as the processes of subjectivation as male or trans within which these somatic events take place. I consider this to be the actual work of the text: that the transformation through the use of testosterone is not described in an authenticating way but is contextualized within a discourse and history of technology focussed on pharmacology. The history of the discourse does not merely frame the narrativization of somatic events. The attributions of meaning – the "signifying acts," to use Butler's terms – are historically, medically and socially so fixed that they are what make somatic events understandable and describable in the first place; they provide a vocabulary. This is what Butler meant by "materialization". We are in the midst of a material-discursive process.

Hairy Arm

In *Pornotopia*, too, in the investigation of the *Playboy* empire, the emergence of juvenile-virile masculinity is emphatically bound to discursive as well as visual, medial and architectural practices. As early as 1987, Teresa de Lauretis used the term "technologies of gender" to describe how visual and linguistic sign systems and representation set in motion the process of becoming a subject, with quite concrete implications for real life. In order to dissolve the sex/gender system the term gender has been emphasized and the ongoing process of doing gender analysed. Unlike Butler, Laureits stresses the importance of technical dispositives and spatial arrangements, which she designates as gender technologies. She argues that gender is shaped by social techniques, cultural representations and practices in the same way that sexuality is implanted in the body. The reason for why there is a final dose of testosterone has to do with these representations, the technologies of gender, which encompass bodies, gestures, spaces and images.[11]

> Yesterday, I took my last dose of T. Today, I can feel the effects. No doubt about it. I'm having a Testo high. One of the first symptoms of testosterone lies in the sensation that the inside of my body is a fibrous and flexible mass that can spread itself through space in any direction [...] But in addition there is slightly more oily skin, sexual excitement, sweat.
> (Preciado 2013, 421)

The end of the experiment with "T" is announced with a scene in a film production company where VD (Virginie Despentes) works. There, VD cuts clips, and Beatriz Preciado (BP) picks them up: "A film production office is a post for the manufacture of masculinity: maximum technology, minimum domestic comfort" (2013, 403). VD masters the entrance into this masculinistic atmosphere with six "film guys" such that she is both "a guy among guys" and a diva who gathers her fans around herself, binding them to herself. (BP is just as much her fan in this scene as I am as a reader.) One of the "film guys" wants to play a casting scene for the other "guys"; the group of men gathers in front of a screen, VD and BP join in without permission. They are gathered to watch the casting of a young woman who undresses bit by bit in front of the camera under the pressure of the demands and touches of the cameraman, who is also the director. Preciado calls the scene, and the protagonist, "Hairy Arm". The cameraman's hairy arm juts into the images from time to time, touching the woman's breasts. And she "imitates the expression of a bitch in a porn movie, the kind of face she appears to have seen a thousand times and has no trouble simulating" (407). But she doesn't get the job; a whiter and purer "girl" is needed.

Hairy Arm is the pharmacopornographic subject par excellence, a subject of each gender, in front of or behind the camera. It is a subject that turns the body into capital, an arrangement of taking possession, of fixed positions and standardized aesthetics. Elsewhere, Preciado called this the "aesthetics of domination of the sexual *ancien régime*" – in response to an open letter by Catherine Deneuve in which she argued that the "culture of seduction" would perish with #MeToo. For Preciado, this argument simply stabilizes a power imbalance by giving it erotic charge, and Paul Beatriz writes this as a "smuggler between two worlds that actually ought not be separated at all, but are kept separated by some as if by a Berlin wall of sex" (2018b).

In *Testo Junkie*, the scene with Hairy Arm in the editing room of a film company marks a turning point: "Will I become a Hairy Arm when I keep on taking testosterone?" (2013, 408). The critical project that *Testo Junkie* understands itself to be aims to unravel the conditions of our pharmacopornographic subjectivity and derail them in a kind of micropolitical act. But self-observation now stands on shaky ground; it is possible that the queer perspective of gender hacking may not be tenable. It could be that the *means of* pharmacopornography cannot be used to *prevail against* pharmacopornography, that the images and texts are too powerful to be reinterpreted, that the perceptions of one's own body under the influence of testosterone are scripted by the narratives made clear in the casting video.

Preciado (2017) resolves to gain entry and takes the administrative-medical route. S/he applies for a change of name and receives a treatment of Testex Prolongatum 250 mg in an injection solution. The point is now "asking to be interpellated as subjects by the ideological apparatus of the very state that excludes them. We ask for recognition (and thus subjugation), so we can then gain agency to invent other forms of subjectivation," as Preciado writes with arguments that are impeccably Butlerian (125). Officials require Preciado's birth certificate to be destroyed, and a new certificate, which is actually false, to be issued. "Thus does law reveal its arbitrary and fictional nature" (131). This is the gap across which the name Beatriz, hiding behind Paul, could be smuggled. The Berlin Wall of sex still stands.

Translation from German into Englisch by Michael Thomas Taylor.

Notes

1 On the notation of *Testo Junkie* as self-experiment *and* research, as personal *and* cool, see Bergermann (2019).
2 Workshops by Diana Torr. Here, see the documentary *A Man For a Day* by Katharina Peters (2012).
3 The history of the "anti-baby pill," with phases of clinical testing in Puerto Rico in colonial contexts, is a separate chapter underscoring that gender issues cannot be examined without considering issues of race (see Precaido 2013, 182–215).
4 The sex/gender system certainly continues to be the basis of institutional women's policy and work to achieve gender equality, in which it then becomes difficult to understand the "strategic essentialism" (Spivak) of affirmative action towards "women" as strategic.
5 The first gender affirming surgeries in fact took place as early as the 1930s. The model of a soul in the wrong body had already been articulated at this point in history. However, in the sexual science of this time no valid distinction between body and soul/mind can be found (see Eder 2010, and Peters 2017).
6 Eve Sedgwick criticized this simplification early on, urging that performativity be understood in more general terms and that the spatial dimension of performativity also be incorporated (1998, 35–67).
7 "[I]s there a way to account for the ability of these norms to materialize the very substance of the human body? [...] If regulatory practices are understood to have a material dimension, how is that materiality theorized?" (Barad 2007, 209).
8 Yet I would argue, by contrast, that it is precisely by excluding psychological dimensions, not to say the unconscious, that the concept of subject remains somewhat insufficiently complex, ultimately working towards a notion of voluntarism, even in Preciado. Here, in my opinion, Butler's work with psychoanalysis and against its normalizing interpretation can offer more convincing conceptions of the subject.

9 In Butler's further defence, it is important to point out that in more recent texts the connections between apparatuses and bodies are quite emphatically considered, and that the concentration on speech acts has been expanded (Butler 2015, 66–98).

10 The term "hormone" is older; it can be traced to Ernest Starling, who introduced it in 1905. Epistemologically, however, gonads and secretions first had to be distinguished from each other.

11 I find Preciado's formulation "technologies of sex," which follows de Lauretis, to be less convincing. I understand the attempt to emphasize the somatic-material level, but does that not again call forth the gender/sex distinction?

References

Barad, Karen. 2007. "Getting Real: Technoscientific Practices and the Materialization of Reality." In *Meeting the Universe Halfways: Quantum Physics and the Entanglement of Matter and Meaning*, 189–222. Durham, NC and London: Duke University Press.

Barthes, Roland. 1975. *The Pleasure of the Text*. Translated by Richard Miller. New York: Farrar, Straus and Giroux, Inc.

Bergermann, Ulrike. 2019. "Test und Testosteron. Medien des Selbstversuchs." In *Differenz und Affirmation: Queer-feministische Perspektiven auf Medialität*, edited by Julia Bee and Nicole Kandioler. Berlin: b_books.

Butler, Judith. 1993. *Bodies That Matter: On the Discursive Limits of "Sex"*. New York and London: Routledge.

Butler, Judith. 2004. *Undoing Gender*. New York and London: Routledge.

Butler, Judith. 2015. *Towards a Performative Theory of Assemblage*. Cambridge, MA and London: Harvard University Press.

De Lauretis, Teresa. 1987. "The Technology of Gender." In *Technologies of Gender. Essays on Theory, Film, and Fiction*, 1–30. Bloomington: Indiana University Press.

Despentes, Virginie. 2015–2017. *Vernon Subutex*. 3 volumes. Paris: Éditions Grasset.

Eder, Sandra. 2010. "The Volatility of Sex. Intersexuality, Gender and Clinical Practice in the 1950s." *Gender & History*, 22 (3): 692–707.

Oudshoorn, Nelly. 1994. *Beyond the Natural Body: An Archeology of Sex Hormones*. London and New York: Routledge.

Peters, Katharina, dir. 2012. *A Man for a Day*. Berlin: Katharina Peters Filmproduktion. Film.

Peters, Kathrin. 2017. "Magic Matters: On Gender Transformation and Early Film." In *Poison and Poisoning in Science, Fiction and Cinema*, edited by Heike Klippel, Bettina Wahrig and Anke Zechner, 119–134. London and New York: Palgrave.

Preciado, Beatriz. 2008. "Pharmaco-pornographic Politics: Towards a New Gender Ecology." In *parallax* 14 (1): 105–117.

Preciado, Beatriz. 2013. *Testo Junkie: Sex, Drugs, and Biopolitics in the Pharmacopornographic Era*. New York: The Feminist Press.

Preciado, Beatriz. 2014. *Pornotopia: An Essay on Playboy's Architecture and Biopolitics*. Cambridge, MA: Zone Books.

Preciado, Paul B. 2017. "My Body Doesn't Exist." In *The documenta 14 Reader*, edited by Quinn Latimer and Adam Szymczyk, 117–134. Munich: Prestel.

Preciado, Paul B. 2018a. *Countersexual Manifesto*. New York: Columbia University Press.

Preciado, Paul B. 2018b. "Letter from a Trans Man to the Old Sexual Regime." In *Texte zur Kunst*. https://www.textezurkunst.de/articles/letter-trans-man-old-sexual-regime-paul-b-preciado/?highlight=preciado.

Runte, Annette. 1996. *Biographische Operationen. Diskurse der Transsexualität*. Munich: Fink.

Sedgwick, Eve. 1998. *Touching Feeling*. Durham, NC and London: Duke University Press.

Spigel, Lynn. 1992. M*ake Room for TV: Television and the Family Ideal in Postwar America*. Chicago, IL: University of Chicago Press.

6 Unthinkable Ecologies in Theatres of the Anthropocene

Ramona Mosse

How can the radical relationality of ecologies be made manifest in narrative and performance? In the 1990s, Amitav Ghosh concluded that there was not yet any aesthetic form that could capture the shaping force of matter on reality: "The truth is we do not yet possess the form that can give the Oil Encounter a literary expression" (Ghosh 1992, 31). This inability stems from the shift in scale, both temporal and spatial, which becomes necessary to capture the operations involved in petrofiction. The story of oil offers a material history that, like no other, has shaped modernity but remains without a narrative (Wilson, Szeman, and Carlson 2017, 5). More specifically, engaging with the history of oil requires an engagement with its "divergent temporalities": human encounters with oil that are characterized by speed (in transportation, for instance) in tandem with material genealogies of fossil fuels that occur on massively enlarged, and hence slowed-down, time scales (Kaposy 2017, 390).

Two decades later, Ghosh has written another study on a missing narrative, focussing not specifically on fossil fuels but rather on how it is possible to narrate and conceptualize the Anthropocene per se. In his *The Great Derangement: Climate Change and the Unthinkable* (2016), Ghosh explores the untold stories of climate change and examines a fundamental interrelation between various forms of oppression and exploitation, ranging from colonialism and patriarchy to natural environments. Ghosh begins with a personal anecdote about encountering a freak tornado in New Delhi and employs the story to show how nature in the shape of a storm becomes reimbued with seemingly supernatural qualities: "Improbable though they might be, these events are neither surreal nor magical. To the contrary, these highly improbable occurrences are overwhelmingly, urgently, astoundingly real" (2016, 27). By describing the experience of a superstorm, Ghosh introduces the idea of the "unthinkable" as a fundamental category within the Anthropocene. *The Great Derangement* thus reframes the paradigmatic spatial and temporal shift necessary to think the Anthropocene. The terminology of the unthinkable even resonates with the scientific definitions of the Anthropocene as a new geological epoch in which "human activities have become so pervasive and profound that they rival the great forces of Nature

DOI: 10.4324/9781003023319-9

and are pushing the Earth into planetary *terra incognita*" (Steffen, Crutzen and McNeill 2007, 614), as the term's inventor, the atmospheric chemist Paul Crutzen, put it. Crutzen's notion of the Anthropocene hones in on a world whose parameters we do not yet have a grasp of. At the same time, Ghosh also makes us aware that the category of the unthinkable stands for both: alternate conceptions of modernity that resist rationalization, and a blind spot when it comes to integrating thinking about natural ecologies with concepts of colonial history. A similar point was also forcefully made by Dipesh Chakrabarty in his seminal essay *The Climate of History: Four Theses*, in which he identified the complexities of species thinking and warned of it "introducing a powerful degree of essentialism in our understanding of humans" (2009, 214) if it is not integrated with an investigation of global capitalism. *The Great Derangement* formulates one such response.

The present chapter, in turn, sets out to answer Ghosh's call for the development of new aesthetic forms by focussing on how theatrical performance puts the complexities of materiality onstage. More specifically, it offers an analysis of Ella Hickson's 2016 *Oil* as an example of the petrofiction that Ghosh was searching for in the 1990s. Hickson's play spans the time frame of the Anthropocene from the 1850s to the present, and interweaves the history of the carbon economy with that of female emancipation and colonial history. Oil itself turns into a medium in Hickson, exemplifying what Stephanie LeMenager has identified as modern media's indebtedness to petroleum energy (2014, 6). In doing so, her play exemplifies how questions of gender, ecology and colonialism come to bear on the performing environments of the Anthropocene. Hickson, alongside other contemporary British playwrights, develops a dramaturgy for exploring what Ghosh identified as the unthinkable relationalities that mark contemporary ecologies.

Unthinkable Relationalities: Or, a Very Short History of Environmental Thinking and Theatre

The deep sense of relationality that characterizes contemporary understandings of both global capitalism and climate catastrophe reaches back to the emergence of ecology as a scientific concept, which is most compellingly illustrated in the work of Alexander von Humboldt. His magnum opus *Cosmos* – begun in 1834 and finally published in several instalments between 1845 and 1847 – set out to compile, chart, measure and imagine a world of interdependencies and connectivities. The five-volume treatise on the material world was not only a conglomeration of Humboldt's extensive travel diaries but also offered a novel conceptual image of the earth as a "wonderful web of organic life" (as quoted in Wulf 2016, 246) in a process of spatial and temporal transformation. Part of the unique acuity of Humboldt's thinking and discovery lies in capturing a fundamental shift in scale, thus preparing the way for the concept of ecology that lies at the heart of this volume. While the zoologist Ernst Haeckel has been credited with first introducing the definition of ecology as a relationality between organisms in his 1866 publication *Generelle Morphologie der Organismen* (*General Morphology of Organisms*), Humboldt's "wonderful web" also captures this kind of ecological conception of the world. *Cosmos*, in particular, was exceptional in its ambition to be comprehensive, and stemmed from what Humboldt himself called "the mad frenzy [...] of representing in a single work the whole material world" (as quoted in Wulf 2016, 246). Humboldt's *Cosmos* was based on a fundamental sense of the possible integration of knowledge, in terms both of this act of representation itself and of the interdisciplinary collaboration between scientists across the spectrum that was necessary to gather such a breadth of material.

Fast-forward to the twenty-first century. The expansive scale with which Humboldt measured the world more than 150 years ago has been replaced by another register: we have moved from the cosmological to the non- or posthuman scale. While Humboldt's cosmological perspective suggests that this planetary scale can be combined with human knowledge practices, the current intellectual debate around the Anthropocene implies a systemic break. Feminist thinkers have been at the forefront of highlighting the biases of modernity's knowledge practices and of the human as a category, while proposing new post- and non-human knowledge formations that interweave gender and ecological thinking. Rosi Braidotti and Donna Haraway are two of the most prolific thinkers when it comes to combining environmental and technological theories with feminism in order to pronounce that "feminism is not a humanism" and instead needs to be decoupled from Enlightenment humanism (Braidotti 2017, 21). Haraway's cyborg in her *Cyborg Manifesto* (1989) lays some of the groundwork for Braidotti's iteration of posthuman feminism. Braidotti also very explicitly references Chakrabarty's "The Climate of History: Four Theses" in her own *Four Theses on Posthuman Feminism (2017)*, where she attacks the lingering human-centred thinking underlying the Anthropocene itself and poses the question: "How can the humanities disciplines – history, literature, philosophy – develop planetary and very long-term perspectives in a geo-centered and not anthropocentric frame?" (Braidotti 2017, 26–27). Braidotti's question encapsulates how much the environmental and climate crisis is, as Lawrence Buell had already put it much earlier, "a crisis of the imagination" (1995, 2). That is to say, the unthinkable pervades contemporary environmental thinking through and through. Haraway notably uses the unthinkable as a core conceptual tool: "Seriously unthinkable. Not available to think with" (Haraway 2016, 30). Haraway most explicitly attacks the concept of the Anthropocene for hiding its evident political and economic dimension behind a pseudo-objective geological narrative:

> Both the Anthropocene and the Capitalocene lend themselves too readily to cynicism, defeatism and self-certain and self-fulfilling predictions, like the "game over, too late" discourse I hear all around me these days, in

> both expert and popular discourses, in which both technotheocratic geoengineering fixes and wallowing in despair seem to coinfect any possible common imagination (56).

Haraway rejects the terminology of the Anthropocene and the Capitalocene on the grounds that they flirt with apocalyptic narratives and reiterate existing power structures based on the distinction of human and non-human worlds rather than foregrounding the ecological entanglements that her alternative approach of the "Chthulucene" sets out to develop: a radical equality of different life forms and their fundamental interrelationality. Haraway instead calls for alternative forms of thinking to rational critique to be established and advocates the creation of new narratives to match the current situation: "It matters which thoughts think thoughts. We must think!" (Haraway, 57).

From Crutzen and Ghosh to Braidotti and Haraway: the unthinkable is a recurring feature in theories of the Anthropocene, connected both to the futurity of scientific projections and to the fatal flaws of the economic and political systems that enable sustained exploitation of the earth's environments. Evolutionist Edward O. Wilson describes a "myopic fog" (1996, 186) that prevents humans from considering factors beyond their lifetime when making plans, and traces how this former evolutionary advantage has turned into a potential disadvantage in the face of the climate crisis. The most prolific theorist of the unthinkable is probably the British philosopher Timothy Morton. While Morton does not use the concept of the unthinkable as explicitly as Ghosh or Haraway, it lies at the core of his concept of "massively distributed" hyperobjects that embody the unthinkable relationalities of contemporary reality, namely, climate change:

> As we enter the time of hyperobjects, nature disappears along with the modern uncertainties that seemed to accompany it. What remains is a vastly more complex situation that is uncanny and intimate at the same time and from which there exists no exit. Thus, the time of the hyperobjects is a time of sincerity, that is a time in which it is impossible to achieve a final distance towards the world.
>
> (Morton 2013, 44)

Here, the density of relationalities itself is a symbol of the breakdown of a crucial critical category: that of the human as a distanced observer and critic. For Morton, it has become impossible to achieve a final distance towards the world. The Anthropocene as the "time of the hyperobject" also posits a categorical challenge to established notions of scientific practice and its method of critical distancing. Only embeddedness and entanglements remain where distance has become impossible. Each of the various theories of the unthinkable makes an explicit demand for alternative concepts of what we might understand by knowledge. One further such alternative can be found in the increased attention to "embodied cognition" with reference to interconnected ecological thinking, which is used as a conceptual framework that "evidence[s] that we use our bodies not only to understand human characters but also for our grasping of the environments that surround them" (Weik von Mossner 2017, 3).

If questions of embodiment and materialities gain in importance in relation to the Anthropocene, theatre as the ultimate art form of embodied cognition seems an obvious choice of tool to capture the world in terms of ecologies, materialities and alternate scales of space and time.

While theatre appears to be a prime laboratory in terms of exploring alternate forms of knowledge creation – being a site of seeing as much as of thinking – it equally seems bogged down by being perhaps the most human-centric of art forms. Drama, after all, is about the representation of human conflict; the modern stage is an inside space that seems sealed off from nature. Una Chaudhuri's near single-handed launch of theatre and performance studies' engagement with ecology in a 1994 article in Yale's *Theater* journal, followed by her book-length study *Staging Place: The Geography of Modern Drama* (1997), privileges instead an account of space, place and environment in the context of an eclectic modern dramatic history, drawing lines between modern drama and the more established discussion of theatrical space in environmental theatre and site-specific performance (1995, 1–20). The idea of place and landscape becomes a formative site of investigation specifically for a theatrically based ecocriticism that spans the canon of modernist to postmodernist texts and performances (Marranca 1996; Chaudhuri and Fuchs 2002; Sullivan 1999) but also extends to the sustainable practical handling of the actual architectural place of the theatre (Fried and May 1994). Bonnie Marranca, in turn, explores ecology and theatre by drawing a genealogy from avant-garde performance practices to contemporary environmentalism (1996, 22–24). Gertrude Stein's redefinition of the "play as a landscape" looms large in this context:

> The landscape has its formation and as after all a play has to have a formation and be in relation one thing to the other thing and as the story is not the thing as anyone is always telling something then the landscape not moving but being always in relation [...] the story is only of importance if you like to hear stories but the relation is there anyway. And of that relation I wanted to make a play and I did, a great number of plays.
> (Stein 1957, 125)

The curious combination of formation and relation allows Stein to shift the core concern of theatre from action to situation, thus creating a landscape out of words no longer tied to the structures of plot but invested in their embodiment in sound and rhythm. The focus on place, space and geography that dominates these theatre ecologies of the 1990s shifted with the rise of the Anthropocene in the 2000s. Thus, in twenty-first-century environmental theatre, the notion of relationality becomes enmeshed with the question of how to place the enlarged planetary scale side by side with the relationalities of global capital. That brings us to Ella Hickson's dramatization of the carbon economy. This also implies a shift from spatial to temporal environments.

Unthinkable Environments I: Gendering Oil

"How does a gathering become a happening?" Anna Tsing asks in *The Arts of Noticing* (2015). She continues:

> One answer is by contamination. We are contaminated by our encounters; they change who we are as we make way for others. As contamination changes world-making projects, mutual worlds – and new directions – may emerge. Everyone carries a history of contamination; purity is not an option.
>
> (Tsing, 27)

Tsing joins Haraway in arguing for a reconceptualization of the world that goes beyond ideas of progress, rationalization and capitalism, and instead involves what she terms new forms of collaboration and precarity. Although Tsing is thinking more generally of the interaction between environmental and social formations, her description of such acts of collaboration as contamination echoes the definition of theatre itself, which in the encounter with others probes and plays with alternative social formations. Theatre as contamination most immediately evokes Artaud's essay on the plague in *The Theatre and Its Double (1938)*, centring the renewal of theatre on liminal experiences over dramatic texts. However, I would like to first investigate a dramatic rendition of contamination as collaboration that addresses the shifting dynamics of human and non-human interaction in the Anthropocene. The contemporary British playwright Ella Hickson constructed a new version of the history play in *Oil*, which premiered in 2016 at the Almeida in London. *Oil* combines the established genre of the family drama (a mother–daughter conflict) with a material history of fossil fuels, roughly covering the time frame of the Anthropocene from the 1850s to the present. The play opens with the encounter between the thick, black liquid that is petroleum and the young woman May, who deserts both her forlorn nineteenth-century Cornish farmstead and her husband in order to follow petroleum on its remarkable march through history. The different scenes move from a Cornwall farmhouse in the 1880s to Tehran in 1908 and Hampstead in the 1970s before leaping into a dystopian future in Kurdistan in 2021 and then ultimately cycling back to the farmhouse. Hickson creates a material history of oil from its discovery as an attractive energy source in the nineteenth century to its increasing dominance throughout the twentieth century. The scenes set in 1908 Tehran highlight the discovery of oil against the backdrop of Iran and Britain's increasing imperial influence over southern Persia following the ratification of the 1907 Anglo-Russian Convention, which divided Persia into different spheres of influence; the scenes in Hampstead explore the 1970s oil crisis and the rise of OPEC; the scenes in Kurdistan depict a future 2021 where oil reserves are dwindling; finally, 2051 offers a vision of a new world and energy order with China at the helm. The material history of oil is thus also always a political history, and in this case a history of British colonialism.

In the play, oil makes its entrance as something immensely seductive that both transfixes May and transforms her from hard-working housewife to time traveller between the ages:

> *A bit of kerosene has spilled on May's finger – she lifts it to her nose and inhales deeply. She loves the smell.*
> Ma Singer: It smells disgusting.
> WW comes in close to May – holds her finger up.
> WW: That was made when the earth started.
> May: Imagine being there then?
> *She turns to him, smiles, transfixed.*
> WW: There are millions of years, right there on the end of your finger.
> May. How can a million years fit on one person's fingers? Magic?
> WW: Near enough.
>
> (Hickson 2016, 18)

While the mother-in-law Ma Singer turns in disgust from the stench of the oil, May grasps its wider significance both as an energy source and, more importantly in this passage, as carrying its own material history within itself: "Millions of years [...] on the end of your finger". Notably, the oil makes its appearance on stage not just through William Whitcomb's (WW) storytelling but as a tangible visual and material presence. The sensory, sensual experiences of the touch (on May's finger) and smell of the oil arguably contaminate May and thus enable her to leave the traditional frame of the interpersonal family drama at the farm and become part of the story of fuel. The sensuality of the experience is further underlined by the parallel interaction with her husband, whom she touches and whose clothes she smells before sneaking out of the house into the night (Hickson 2016, 26). In a sense, May's departure is staged as a betrayal of her husband: not with another person, but with petroleum itself. Equally, the staging itself gives oil a dramaturgical presence, as the oil literally brings a different quality of light to the stage: the set design moves from a realistic chamber play with reduced lighting to a multidimensional, intermedial space, in which the combination of stage and screen extends the possibilities of performing how fossil fuel meets humankind.

Structurally, Hickson employs a set of poetic inter-scenes that allow May to travel through time. Throughout, her own career mirrors the rise of oil as the dominant energy source of the twentieth century, so that she ends up as a fossil fuel executive. However, it is the ambivalence of May in her various symbolic functions throughout the play that fuels the theatrical and intellectual experience of *Oil*. For example, leaving her husband and the narrow confines of the Cornwall farm in the first scene is staged decisively as an act of female liberation from the confines of patriarchal social and economic structures that tie family and land together by way of tradition. Notably, Ma Singer bids May a sarcastic farewell – "If you are stepping out – make sure that door is closed behind you" (2016, 26) – thus inverting Nora's infamous slamming of the door at the end of Ibsen's *A Doll's House*, which sets Nora free from the strictures of the household. May, in contrast, opens the door and steps out into the future in a much more literal sense.

Patrick Lonergan sees a close link between Hickson's experimentation with temporal structures in the play and Timothy Morton's *Dark Ecologies*. He comes to the conclusion that "Hickson has created a play that confounds our understanding of both the real and the unreal" and "shows the limitations of both categories" (Lonergan 2020, 44). In Lonergan's reading, Morton's evocation of hyperobjects imprints itself on the dramaturgy of *Oil*. While Lonergan is right to highlight the experimentation with different levels of reality, he equates theatre with a dramatic realism that does not sufficiently account for alternative strands of theatrical aesthetics. Neither theatre nor drama for that matter is exclusively or necessarily tied to realism in this way. In fact, May's departure from the farm is also a conscious farewell to the conventions of dramatic realism. At the same time, Hickson retains some of the conventions of the family drama, most strikingly in the intertwining of gender and economic relations. This becomes particularly evident in Part 2, set in Tehran, which finds May working as a maidservant at a colonial mansion with her daughter Amy, who has now been born and is eight years old. It is a party scene that revolves around the theme of consumption. The excess of the food and drink for the party guests resonates with the background: "On the horizon, the oil wells of Iran. Working endlessly" (Hickson 2016, 27). This signifies the insatiable desire for more oil. Meanwhile, May and Amy struggle with hunger and lack of pay, and are themselves objects of desire to be consumed or owned:

> May: Officer Samuel, the girl is right there.
>
> *Samuel goes to grab May. She tries to* push *him off but he really goes for it – runs his hand up her leg.*
>
> Amy *(with her mouth full):* Mummy.
>
> *May suddenly spots Amy who had stuffed her face entirely with the Turkish*
>
> *Delight – it's horrible somehow, she looks like she is going to be sick.*

May: Amy!
Samuel: Leave her, she's fine – for a moment, just leave her.
May is trying to wiggle free from Samuel to get to Amy but he is holding tight.
Ana goes over to Amy – gets there first, tries to help. May arrives and pushes
her off.
She's my child! I'll deal with it – she's my child!

(Hickson, 49–50)

The lines of possession cut across several divides and operate generationally, ethnically, sexually, economically. May's struggle with Samuel, who represents the British colonial oil interests, is a trading deal, in which her body is exchanged for the well-being of her and her daughter. Meanwhile, May's objectification does not remain an isolated instance, but instead becomes mirrored in the mother–daughter relationship and the concurrent rivalry between May and the Persian maid Ana, whom Amy claims to prefer as a mother. The struggle around erotic desires is substituted by maternal claims, as May stakes her ownership of Amy, which, in turn, reflects the colonial hierarchies between the British and Persian subjects encapsulated by the May–Ana standoff. During this struggle Amy sits under the table, unable to stop consuming the Turkish delights that do not belong to her. Once more, another version of ownership and a lack thereof dominate the dynamics of the scene. This moment of excessive consumption also gestures again to the entanglement of imperial powers – specifically, the British and Russian empires vying for control of Persian natural resources in the early 1900s – with the material history of the carbon economy. Ghosh describes the Great Derangement (after which his book is named) as follows: "We are mired in the Great Derangement: our lives and our choices are enframed in a pattern of history that seems to leave us nowhere to turn but to our self-annihilation" (2016, 111).

None of the relationalities between the play's characters are particularly original in and of themselves (the generational struggle between mother and daughter, the patriarchal gender relationships, the suppression of the subaltern or the metaphor of consumption that runs throughout the play), yet their cumulation as a stand-in for the history of oil opens up novel ambivalences and tensions for theatre audiences. The centrifugal point of ambivalence is May herself. She is fearlessly progressive and her story as a single mother is a marker of female emancipation but who simultaneously becomes an icon for exploitation, opportunism and capitalism, as she rises to the rank of oil industry executive. Hickson highlights how female emancipation in the West was highly dependent on the material conditions of technological advancement, encapsulated by the oil industry; her ambivalent figure of May thus creates a counterpoint to ecofeminist philosophies and their equation of feminism with environmental awareness, advocacy and a mirroring of feminism in the concept of Mother Nature. Instead, female emancipation becomes possible only within the context of advancing technological development. Likewise, in the gender ecology of the play, any binary between male and female is undone as the decisive conflicts and decisions that drive the play exclusively involve the female protagonists. Ultimately, *Oil* exchanges dramatic realism for a secularized version of medieval allegory, since the connectivities and relationalities in the play no longer stand for themselves but as an iconography for the invisible intertwining of capital and environment. Within that allegory, the more abstract oppression of Western colonialism is mapped onto female bodies that struggle to untether themselves from these forms of domination.

Unthinkable Environments II: Scales of Climate Science

While *Oil* ends in 2051, British playwright Duncan Macmillan's portrayal climate science in *2071* takes the audience even further into the future, though in a very different dramaturgical context that questions the efficacy of dramatic theatre in addressing climate change. While Hickson's play translates the history of fossil fuel into the confines of a family drama, albeit with a twist, Macmillan instead gives a verbatim staging of a climate science lecture by Chris Rapley, Antarctic scholar and former director of the Science Museum in London. The play *2071* is co-written by Rapley and Macmillan and opened simultaneously at the Royal Court in London and the Schauspielhaus in Hamburg in the run-up to the Paris climate conference in 2015. It was directed by Katie Mitchell and represents a continuation of Mitchell and Macmillan's work together to explore climate change through theatre, most famously in their Schaubühne Berlin production of *Lungs* (2011), in which the actors themselves generated the energy for the stage lighting by cycling during the play.[1]

Despite the obvious dramaturgical and structural differences between verbatim and dramatic theatre, I would like to begin with a striking parallel symbolism between Hickson's drama and Macmillan's performance lecture. At the heart of Rapley's lecture is a scene that resonates intriguingly with May's first encounter with petroleum:

> As the Director of the British Antarctic Survey on one of my Antarctic trips in 2002, I visited the European drill site at a place called Dome C. I watched as a 5m section of ice core, nearly half a million years old, was extracted from a depth just under 3km. It took an hour to lower the drill, a few minutes to drill the core section, and an hour to winch it up to the surface. There are offcuts, small chunks of the core, which are not useful to science. I picked a piece up. As a scientist I try to remain objective and dispassionate. But here I was, in a part of the world that had fascinated me since I looked at that area marked "Region unknown

> to Man" as a child, holding a piece of ice that had not seen the light of day since before the dawn of mankind. I listened to the air bubbles pop and crackle as the ice melted from the heat of my hand. I breathed the air coming out of it, air that was trapped at the time of freezing.
>
> (Macmillan 2016, 203)

Much like May, who takes in the odour of the petroleum on her fingertip, Rapley not only holds ancient ice but also breathes it in, a gesture that vividly symbolizes the permeability and relationality between self and environment. It thus turns into another act of contamination, in Tsing's sense of the term, that is a moment of transformation. Despite Rapley's initial attempt to maintain his distanced objectivity as a scientist, he goes on to reveal that holding the offcuts of ice, which are essentially useless, provided him with a moment of intense passion, similar to the seduction experienced by May and ultimately echoing Morton's previous verdict that objective distance has become impossible in the age of hyperobjects. As a gesture, it also encapsulates the fluid negotiation in *2071* between the representation of relationality through data and embodiment. The abstraction of data meets and intertwines in the play with the concreteness of embodiment and ultimately suggests that these two poles can no longer be separated but are both integral parts of knowledge processes. The authors' opening note in *2071* highlights a distinctive tension between scientific data and theatrical spectacle, which might "oversimplify the issues and compromise the integrity of the data" but equally acknowledges climate science as "a complicated and emotive topic" (Macmillan 2016, 191). This classification as an emotive rather than a purely data-driven or rational enterprise is essential, since it can offer the beginning of an answer to Rosi Braidotti's opening question about the role of the humanities in the age of the Anthropocene. Moreover, *2071* offers an interesting inversion of the passage back from the performance of self to performance as role-play. Written for and as an adaptation of Rapley's scientific lecture, the opening note to the play suggests that "as a performance text, it can be performed by one or more people, of any age, gender, class, or ethnicity" (Macmillan 2016, 191), thus broadening out beyond the figure of Rapley himself. As a performance text, the *2071* lecture combines the abstract complexity of climate science with its affective dimension through moments like Rapley holding the prehistoric ice. Even more importantly, the text makes the position of the scientist itself accessible by allowing it to be inhabited by anyone and everyone. In fact, taking up the role of scientist and giving voice to their findings becomes a necessity for each audience member: "For the [Paris] agreement to work, we shall have to play our part" (Macmillan 2016, 191) – and that means inhabiting the world that Rapley evokes in his lecture. Thus, the most radical gesture in Macmillan's and Rapley's performance text might not be the passion invoked in experiencing the ice melting on a fingertip but instead the willingness to move from the self-performance of Rapley to a general call to "play our part" *as* Rapley, thereby acknowledging and voicing our own deep entanglement in climate science. Ultimately, this verbatim play turns into an example of what Andy Lavender has termed new "theatres of engagement" (2016, 23), which stand out by being "something other than an encounter between actors, or between actors and audience. There is no longer a separation between the space of performance and that of spectatorship. Scenic space is inhabited" (10). In other words, as participants everyone is united in a scenic space marked by relationality.

By Way of Conclusion: Ecologies as Sites of Relationality

The common relationality explored in the otherwise very different dramaturgies of these theatrical texts returns us to Gertrude Stein's concept of a play as a landscape, something wholly situated and relational. However, the environments that Macmillan, Rapley and Hickson create gain their level of unthinkability in large part from their temporal rather than their spatial dimension. In other words, material environments are experienced through a sense of temporal shifting and dispersion: the fossil fuel creates a temporality that takes the audience back to prehistoric times and catapults them forward through modernity's interactions with oil, while on Macmillan's stage water unfolds its own material history – one that extends beyond the human, both into the past and into the future. May walks out of a specific setting to become a time traveller while Rapley sits fixed in his chair, as if to denounce or counteract the theatricality of the space he inhabits and instead has the audience imagine breathing in the air from half a million years ago. Both scenes establish the materiality of water and oil respectively as a trace of temporal scaling, when half a million or a million years suddenly rest on someone's fingertip: a thought both immediate and unthinkable. The temporal scales are not only, like Morton's hyperobjects, vastly distributed, they are also looped and twisted rather than solely monodirectional. Experiencing environmental relationality becomes embodied for Rapley and May in what Rebecca Schneider has called "the fold" that creates a "tangle of explicit theatricality and time" (2011, 6) – the experience of a simultaneity of present, past and future beyond the human moment. Temporal scale becomes embodied both as physical and written verbal gesture. In this suspended moment of thinking a history without the human and simultaneously being incapable of totally eradicating oneself from that thought lies the poetic dimension of these plays about different ecologies. Equally, to play not with the fold of time but the spatial fold of ecologies is to realize that the engagement with these ecologies and their alternative knowledge bases brings with it a necessary reassessment of the category of the human both as a species and as a multitude of splintered and incommensurable gendered identities.

Note

1 For a trailer and short description of the production, see the Schaubühne Berlin website: https://www.schaubuehne.de/en/produktionen/atmen.html?m=316.

References

Braidotti, Rosi. 2017. "Four Theses on Posthuman Feminism." In *Anthropocene Feminism*, edited by Richard Grusin, 21–47. Minneapolis: Minnesota University Press.

Buell, Lawrence. 1995. *The Environmental Imagination*. Cambridge, MA: Harvard University Press.

Chakrabarty, Dipesh. 2009. "The Climate of History: Four Theses." *Critical Inquiry* 35 (2): 197–222.

Chaudhuri, Una. 1994. "'There Must Be a Lot of Fish in that Lake': Toward an Ecological Theater." *Theater* 24 (1): 23–31. https://doi.org/10.1215/01610775-25-1-23.

Chaudhuri, Una. 1997. *Staging Place: The Geography of Modern Drama*. Ann Arbor: University of Michigan Press.

Fried, Larry, and Theresa J. May. 1994. *Greening Up Our Houses: A Guide to a More Ecologically Sound Theatre*. New York: Drama Books.

Fuchs, Elinor, and Una Chaudhuri. 2002. *Land/Scape/Theatre*. Ann Arbor: University of Michigan Press.

Ghosh, Amitav. 1992. "Petrofiction." *The New Republic* 206 (9): 29–34.

Ghosh, Amitav. 2016. *The Great Derangement: Climate Change and the Unthinkable*. Chicago, IL: University of Chicago Press.

Haraway, Donna. 2016. *Staying with the Trouble: Making Kin in the Chthulucene*. Durham, NC and London: Duke University Press.

Hickson, Ella. 2016. *Oil*. London: Nick Hern Books.

Kaposy, Tim. 2017. "Petroleum's Long Duree: Writing Oil's Temporalities into History". In *Petrocultures: Oil, Politics, Culture*, edited by Sheena Wilson, Adam Carlson and Imre Szeman, 389–406. Montreal: McGill-Queen's University Press.

Lavender, Andy. 2016. *Performance in the 21st Century: Theatres of Engagement*. London and New York: Routledge.

LeMenager, Stephanie. 2014. *Living Oil: Petroleum Culture in the American Century*. Oxford: Oxford University Press.

Lonergan, Patrick. 2020. "A Twisted, Looping Form." *Performance Research* 25 (2): 38–44. https://doi.org/10.1080/13528165.2020.1752575.

Macmillan, Duncan. 2016. "2071." In *Plays One*, 190–216. London: Oberon Books.

Marranca, Bonnie. 1996. *Ecologies of Theatre: Essays at the Century Turning*. Baltimore, MD: Johns Hopkins University Press.

Morton, Timothy. 2013. "Poisoned Ground: Art and Philosophy in the Time of Hyperobjects." *Symploke* 21 (1–2): 37–50.

Oswald, Alice. 2020. "Nobody." *The New Yorker*, August 17. https://www.newyorker.com/magazine/2020/08/24/alice-oswalds-homeric-mood.

Schneider, Rebecca. 2011. *Performing Remains: Art and War in Times of Theatrical Reenactment*. London and New York: Routledge.

Schuster, Joshua. (2015). *The Ecology of Modernism: American Environments and Avant-Garde Poetics*. Tuscaloosa: University of Alabama Press.

Steffen, Will, Paul Crutzen and John McNeill. 2007. "The Anthropocene: Are Humans Now Overwhelming the Great Forces of Nature?" *AMBIO: A Journal of the Human Environment* 36 (8): 614–621.

Stein, Gertrude. 1957. *Lectures in America.* Boston, MA: Beacon Press.

Sullivan, Garrett. 1999. *The Drama of Landscape.* Stanford, CA: Stanford University Press.

Tsing, Anna. 2015. *The Arts of Noticing the Mushroom at the End of the World: On the Possibility of Life in Capitalist Ruins.* Princeton, NJ: Princeton University Press.

Weik von Mossner, Alexa. 2017. *Affective Ecologies: Empathy, Emotion and Environmental Narrative.* Columbus: Ohio State University Press.

Wilson, Edmund. 1996. *In Search of Nature.* London: Penguin.

Wulf, Andrea. 2016. *The Invention of Nature: The Adventures of Alexander von Humboldt, the Lost Hero of Science.* London: John Murray Publishers.

Part 3

Spaces

Landscapes and Architectures of Power and Imagination

7 Gender, Nature, Nonhuman Animal

Bird People (2014) and the Proliferation of Difference in Cinema

Sabine Nessel

Introduction

An open book, stood upright against a white background. The pages and cover are damaged. The edges of the paper are ragged, with small pieces missing; it looks as if they have been cut out. The title, printed in big letters, "Die Welt in der wir leben" (the German edition of the *Life Magazine* publication *The World We Live In*) can no longer be completely made out. What at first glance appears like a simple visualization of themes of crisis in human–nature relations that are discussed under headings such as "the Anthropocene" is the result of a concrete process of production. It is a recent work by the interspecies collective CMUK, comprising two human agents, the artist duo Hörner/Antlfinger, and two animal agents, Clara and Karl, a pair of African grey parrots (Figure 7.1).[1]

The artists Ute Hörner and Mathias Antlfinger gave the grey parrots one of the most famous books of the twentieth century to work on. Originally published in the 1950s, *The World We Live In* is a now-outdated account of the history of evolution that uses the concepts "world" and "we" in its title in a straightforward sense. The grey parrots worked on a copy of the book with their beaks and claws (Ullrich 2016) or, if you will, subjected it to an animal re-reading. The book object combines deconstructive processes with a mode of production centred on interspecies collaboration, in which practices of writing, inscribing, reading and quoting are combined with ones of gnawing, scratching and animalistic reworking. What does this opening example show us? The interspecies collective's work can be understood, first, as a critique of an account of the earth's history that enjoyed popularity in the 1950s; second, the artistic authorship is special because of the crucial role played by both animal and human agency, something also indicated by the initials CMUK (Clara, Mathias, Ute, Karl) that make up the art collective's name. We might even say that, as animal protagonists, the grey parrots Clara and Karl make a contribution that can no longer be unambiguously assigned to the side of either nature or culture, but instead prompts us to ask just how the relation between nature and culture is constituted and is to be described in this case. The choice of names, Clara and Karl, is also

DOI: 10.4324/9781003023319-11

Figure 7.1 Book object *The World We Live In* (CMUK Interspecies Collective, 2016) Photo: Hörner/Antlfinger.

interesting, as it codes the parrots as gendered. The book object *The World We Live In* thus combines a critique of a certain account of the earth's history with an interrogation of differences, such as that between nature and culture, between animal and human or between different genders.

In recent discourses such as new materialism, cultural animal studies,[2] the environmental humanities and the discourse of the Anthropocene, environment and nature are understood as complex configurations at the intersection between agents, global states of affairs and cultures (Haraway 2016b; Chakrabarty 2017). The concepts of "gender", "nature" and "nonhuman animals" form part of these discourses, and what they have in common is that they make reference, each in their own specific way, to questions of difference. The history and theory of systems that produce gender difference have attracted attention in the human, social and natural sciences alike, and have had an especially marked influence on newer disciplines such as film and media studies. At the same time, notions of difference have sometimes generated blind spots, concealments and exclusions, whereby one difference is overlooked in favour of another (see Ngai 2006; Nessel 2019). This chapter focusses on concepts that attest to a proliferation of difference. My hypothesis is that within the proliferation of difference we can discern the historicity of the idea of individual difference. Following in the footsteps of Kimberlé Crenshaw, theories of intersectionality have fruitfully drawn attention to the relevant connections between forms of social inequality and discrimination along axes of gender, ethnicity and sexuality (Meyer 2017). In order to meaningfully examine the proliferation of difference in relation to arts, media and film, it is necessary to adopt a perspective in which aesthetic, technical and audio-visual aspects of media forms and modes of production are fundamentally integrated.

This chapter shall attempt to show how nature relations can be analysed through a film and media studies lens informed by gender and cultural animal studies. It is divided into three sections. The first section begins by discussing Donna Haraway's "figures" and Jacques Derrida's "animots", two key concepts in feminist discourse and cultural animal studies respectively, and considers how they can be connected to studies of the proliferation of difference in arts and media. The second section is a case study of Pascale Ferran's film *Bird People*, in which questions of anthropological difference figure prominently, alongside questions of gender and class difference. The third section then proposes a perspective shift on nature relations that opens up an alternative reading of the film. This perspective shift focusses on a hub of global capitalism, Charles de Gaulle Airport in Paris, and on selected material–semiotic nodes or knots that attest to a proliferation of difference.

Possible Points of Connection: Concepts of Difference in Haraway and Derrida

If we are enquiring into intersections of difference, there are two concepts that suggest possible points of connection with arts and media. The first originates in history of science, the second in philosophy, and they have prompted debate in gender studies and cultural animal studies respectively. Donna Haraway contributed to feminist discourse in the 1980s with her text "A Cyborg Manifesto: Science, Technology, and Socialist-Feminism in the Late Twentieth Century" (Haraway 2016a, originally published 1985). Her complex deconstructive methods, operating at the intersection between images, texts, narratives and discourses, have been highly influential in European and Anglo-American history of science as well as in artistic and popular culture. The Cyborg Manifesto argues that dichotomies of mind and body, animal and human, organism and machine, public and private, nature and culture, man and woman, are "in question ideologically", and focusses on the analysis of the conditions of possibility of the culturo-historical oppositions of male–female and human–machine. According to Haraway,

> certain dualisms have been persistent in Western traditions; they have all been systemic to the logics and practices of domination of women, people of color, nature, workers, animals – in short, domination of all constituted as others, whose task is to mirror the self.
>
> (Haraway 1985, 59)

She argues that the culture of high technology poses a challenge to these dualisms: in the relation between human and machine, it is unclear who or what creates and who or what is created. What is mind and what is body in machines has become undecidable. While the Cyborg Manifesto attempts to lead us out of the labyrinth of dualisms using the metaphor of "the cyborg", in *When Species Meet* (2008) it is Haraway's pet dog that occupies a central place. The latter work raises two principal questions. Firstly, "Whom and what do I touch when I touch my dog?"; secondly, "How is 'becoming with' a practice of becoming worldly?" (2008, 3) The Other appears in Haraway's study in the guise of her dog Ms Cayenne Pepper, who, like the cyborg in her earlier essay, can be understood metaphorically. The cyborg and Ms Cayenne Pepper, as well as Coyote the trickster, are prime examples of what Haraway labels "figures":

> Figures are not representations or didactic illustrations, but rather material–semiotic nodes or knots in which diverse bodies and meanings coshape one another. For me, figures have always been where the biological and literary or artistic come together with all of the force of lived reality.
>
> (Haraway 2008, 4)

While Haraway's concept of "figures" has been widely discussed within feminist and gender discourses, Derrida's "animots" is a key concept in European cultural animal studies. In *L'animal que donc je suis* (*The Animal That Therefore I Am*) Derrida distinguishes between two types of discourse about animals. On the one hand, there are "texts signed by people who have no doubt seen, observed, analyzed, reflected on the animal, but who have never been *seen seen* by the animal" (Derrida 2008, 13). On the other, there are "signatories" who "admit to taking upon themselves the address that an animal addresses to them, before even having the time or the power to take themselves off [*s'y dérober*]" (14, insertion in square brackets in original). Derrida classes his own discourse on animals as belonging to the second group. In the text, he appears in the guise of a philosopher who, on his way to the bathroom one morning, encounters his cat. He feels like it is watching him and reacts with shame (which prompts him to put on a dressing gown). "It is as if I were ashamed, therefore, naked in front of this cat, but also ashamed for being ashamed" (4). While Heidegger (who Derrida assigns to the first group of discourses) holds that pets "live" with us, while we merely "are" with them, thereby underscoring the traditional opposition between human and animal in which "the animal" is conceived as a species representative, Derrida chooses to include in his argument his pet cat and the shame he feels before its gaze. The appearance of Derrida the philosopher and his cat in his own text is no mere joke, but forms the basis for a deconstructive reading of philosophical discourses on animals. It opens up a perspective on animals that diverges from their traditional place in Western metaphysics. Instead of using "the animal" as a designation of being, Derrida deploys the neologism *animots*, a combination of *l'animal* (animal) and *le mot* (word). This expression shares a formal similarity with *différance*; when spoken, *animots* is audibly indistinguishable from *animaux*, the French plural form of "animals", thereby displaying the fact that we only synthesize animals "into a unity through our language" (Wild 2008, 205).

What points of connection do Haraway's "figures" and Derrida's "animots" offer for investigations through a film and media studies lens? While Haraway's notion of "figures" incorporates visual and aesthetic dimensions and emphasizes the interconnectedness of biology, art and life as "material–semiotic nodes or knots" (Haraway 2008, 4), Derrida's term "animots" marks the mediated dimension of the expression "the animal" in spoken language. The animal as a species representative is treated not as a firmly established notion or mere descriptor, but is understood rather in terms of its genesis as a linguistically produced dimension and hence as mediated. By contrast with the cat's gaze that figures in his argument, Derrida's study makes no mention of animal voices or animal articulation. The dimension of sound comes in elsewhere, namely in the spelling of *animots*. Derrida draws attention to the inaudible nature of the difference between *animaux* and *animots*, which becomes apparent only in writing. The acoustic dimension is implicit, understood anthropologically by reference to the human voice; the central argument is based on a pattern of sound.

One good example of how "figures" and "animots" can make visible a film's perspective on intersections of difference is a scene in Robert Bresson's *Au hasard Balthazar* (1966) that takes place at a circus. Even before the circus animals are shown one after the other on screen, their voices are audible as a cacophony of animal noises. The film draws a connection between Balthazar the donkey, the main character, and the circus animals in cages both through the soundtrack and through the visual means of reverse cuts, making it seem as though the circus animals are talking to Balthazar as he passes mutely through. The impression that the animals are having a "conversation" (that we cannot understand) is reinforced by showing the exchange of gazes between them. However, the voice of Balthazar is prominently exposed throughout the film. According to one report on the film's production, filming the shots of the donkey braying was a complicated procedure (Wiazemsky 2009, 135–136). The main problem was getting the donkey to bray at the right moment. This problem was solved by the trick of playing the donkey a recording of its own voice that had been made earlier. The donkey on the film set thought it was hearing another donkey and brayed in reply. The shot of Balthazar the donkey braying was predicated on the technical possibility of reproducing its voice.

While Derrida coined his concept of *différance* by reference to the history of writing in Western philosophy, *animots* was coined based on an episode from lived experience (going to the bathroom). Aside from the triad of biology, literature and life that underpins the neologism *animots* and that connects Derrida's and Haraway's concepts, Derrida seeks to deconstruct the ontology of "the animal", and hence remains at the level of language and discourse. As Haraway rightly observes with regard to Derrida's coining of *animots*, it is not enough "to speak about animals or of animals; the goal must be to act in concert with animals. For animals, that means they are kept in play as active agents" (Haraway's argument as paraphrased by Borgards 2012, 43). Haraway's call for humans to act in concert with animals is applied in practice in the scene from Bresson's *Au hasard balthazar* that was described earlier; the film crew on the set studied the donkey's behaviour and set things up in such a way that the donkey would bray at the right moment.

A real-life episode[3] in which Derrida is watched by his cat as he walks to the bathroom naked prompts a shift of focus to the philosopher's exchange of gazes with the animal.[4] A business executive's exchange of gazes with an animal, in this case a sparrow, is another story. That story is depicted in complex, spectacular fashion in the feature film *Bird People* (2014).

(Non)Human Animal: Intersections of Difference in *Bird People*

Bird People, a film by French director and screenwriter Pascale Ferran, premiered at the Cannes Film Festival on 19 May 2014 in the *Un Certain Regard* section. The film is based around a dual story of transformation. The first transformation is depicted by the character of Silicon Valley executive Gary Newman (Josh Charles), who after a meeting suffers a breakdown, quits his job and leaves his family. The second transformation story centres on the character of Audrey Camuzet (Anaïs Demoustier), a student working at a hotel who during the film turns into a sparrow. The film is mainly set at the Hilton Hotel at Charles de Gaulle Airport in Paris. *Bird People* documents contemporary social conditions in the age of digitalization and globalization, depicting "flexible people" (Sennett 1998) in "non-places" (Augé 1992). It shows urban spaces as spaces of transfer, explores the world of work and brings diverse modes of existence into focus: on the metro, at the train station, at the airport. In the opening scene, Gary's and Audrey's separate journeys to, and eventual arrival at, the hotel are shown: travelling on the train, passing by coffee machines, surrounded by passers-by looking at their phones, getting caught up in the throng of rush-hour commuters, taking a taxi and so on. This opening is followed by two chapters marked by intertitles: "Gary" and then 43 minutes later "Audrey".

The car in front of the taxi that is taking Silicon Valley executive Gary down the motorway to his hotel after a meeting turns on its hazard lights. Immediately ahead, the motorway narrows to a single lane, and the reason for the lane closure – a serious accident – comes into view. Gary's gaze now focusses on the scene of the fatal accident, the wrecked cars surrounded by broken glass. The scene on the motorway is shown in remarkable detail, but only acquires its full significance later on. Up until this point, everything seems to be business as usual: stressful meetings, highly demanding work and intense deadline pressure are all everyday situations in, and fundamental pillars of, the (primarily) Western neoliberal economy, which encompasses every sphere of life from business to culture to academia. But the motorway accident also foreshadows what is to come. The drama of the accident resurfaces in psychosomatic form a little later on when Gary has a panic attack that transforms his life and sets the film's plot in motion. When he can't stand the feelings of anxiety any more, he runs down from his hotel room and out the front doors, watched by the night porter Simon (Roschdy Zem). Gary has gone from being observer, watching the dramatic images of the accident from a safe distance, to observed, racing out of the hotel foyer as if he were being remote-controlled. Gary's transformation in the wake of his panic attack is shown in minute detail, as he quits his job and leaves his family.

The second main strand driving the action of the film is Audrey's transformation into a sparrow. While Gary's transformation is depicted realistically, with scenes showing spaces of transit and the ways that senior managers talk and act, Audrey's transformation occurs with virtually no advance warning in a realm located somewhere between everyday reality and fantasy. The film's narrative explains the causes of Gary's transformation, but there is something abrupt and astonishing about Audrey's. It happens when she goes up onto the hotel roof after a power cut. Her metamorphosis into a sparrow comes as a surprise not just to the audience but also to Audrey herself. Like in the Superman films, her startlement at the transformation is accompanied by self-commentary and an exploration of the new possibilities it brings with it. The first flight that Audrey takes in her sparrow body bears the hallmarks of a test flight, at the end of which she lands on a bin outside the Hilton's back doors. She also practises flying in a straight line and playfully pretends to race against a big passenger plane as it takes off.

Audrey's transformation into a sparrow brings the idea of anthropological difference, of what separates animal from human, to the fore, which in Western philosophy has persistently been defined by reference to language. From Aristotle to Descartes, from Kant to Heidegger, Lévinas and Lacan, the animal is held to be "deprived of language" (Derrida 2008, 32). While the philosophical tradition denies animals the power of speech, there are plenty of examples of speaking animals in cultural works throughout history, such as the Old Testament, Goethe's *Reineke Fuchs* (1794), Kafka's *A Report to an Academy* (1917), where Red Peter the ape tells the story of his transformation – and of course *Bird People*.

After landing on the bin at the back of the Hilton, Audrey (as a sparrow) listens in on the conversation between the night porter Simon and a colleague about their living conditions. When they're suddenly interrupted by a mobile phone ringing, Audrey-as-a-sparrow's gaze meets that of Simon. Only later in the film, when Audrey (as a sparrow) follows Simon unobserved into a wooded area nearby does it become clear that he concealed something in the conversation with his colleague. Contrary to what he claimed, he actually sleeps in the back of his car at night. He and other homeless people have set up temporary night quarters in a car park in the woods. The exchange of gazes between Simon and Audrey (as a sparrow) can be connected to Derrida's concept of "animots" and Haraway's material–semiotic nodes or knots, in which differences such as gender, class, colonialism and migration intersect. While for Derrida it was being seen naked by his cat that triggered feelings of shame, for Simon it is the sparrow's gaze. The philosophical argument has been brought down to earth. Simon's shame at his homelessness and precarious circumstances becomes palpable in the proliferation of difference. Notably, Simon is played by the French actor Roschdy Zem, who is the son of Moroccan immigrants (and in his own films usually portrays immigrants of Moroccan or Algerian heritage). This situates Charles de Gaulle Airport not just as an icon of global capitalism but also within the context of France's colonial history.

Another exchange of gazes, which the film elaborately builds up to and marks out as powerfully significant, occurs on a moving walkway at the airport. That is where Gary and Audrey (as a sparrow) meet, the one ambling along, the other flying around the spacious terminal. This time, the set-up is far more complex, involving two parallel walkways moving in opposite directions and the shot–reverse shot principle. We start with Gary's point of view, then shift a moment later to Audrey (as a sparrow) and then back again. The shot–reverse shot montage presents the strange meeting of species as an encounter between equals, (almost) eye to eye. Becoming-animal is shown neither as a transition to pure corporeality nor as a loss of intelligibility; rather, textuality and writing play a central role. However, maps, diaries and pieces of paper are perused not from a physical distance but from a position right in the midst of things. When Audrey (as a sparrow) finds a map of France laid out on the bed in Gary's room, she flies up to it and "reads" it by hopping around (Figure 7.2).

Eye movements are replaced by the movements of Audrey's (sparrow) body as she/it explores the map, reading the names of the cities out loud: Biarritz, San Sebastián, Perpignan.

During her flight around the airport, Audrey is uncertain in her new existence as a sparrow (perhaps precisely because it is so astonishingly free). At the start, we can hear her voice exclaiming "allez!" and "vite, vite, vite", which codes the bird's perspective on the airport as female. Only after that does David Bowie's "Space Oddity" kick in. The song was inspired by Apollo 11, the first mission to land humans on the moon. It takes the form of a dialogue between the fictional astronaut "Major Tom" and "Ground Control" back on earth. The song's cult status can be attributed to the old dream of flying to the moon, which Georges Méliès brought to the cinema screen in 1902 and that came true in 1969 with the first lunar landing – and was then euphorically taken up by art and popular culture in the years that followed. As the song begins, the film leaps into another galaxy and the (male-connoted) first moon flight intermingles with the (female-connoted) sparrow flight. The cosmic-seeming relation between the two flights is rooted in a complex *mélange* that interweaves discourses of gender difference, anthropological difference and difference between nature and culture/technology.

Figure 7.2 Audrey (as a sparrow) on a map (*Bird People*, dir. Pascale Ferran, 2014).

In terms of production, Audrey's becoming-animal was the result of complex processes that included animal training and visual effects added in post-production. The sparrow that Audrey transforms into was created using several different methods, and variously appears as a full 3D bird or as a 2D or 2D/3D bird based on live bird shots. Shots of real sparrows before the camera are a creaturely trace, an old-fashioned way of redeeming physical reality, which here does not stand in opposition to the digital, the fictional, the synthetic, but rather forms a material–semiotic knot or node akin to Haraway's figure of the cyborg. It is notable that the director worked with a group of sparrows over an extended period of time to produce the live bird shots,[5] studying their behaviour and training them in a manner reminiscent of production methods that were prevalent in early cinema, when actors like Berthe Dagmar and Nell Shipman appeared before the camera with their own pets – very much in line with Haraway's call for humans to act in concert with animals (see Förster 2012). Audrey's transformation, which transcends the film's narrative level and traditional ways of depicting spaces of digitalization and globalization, will in the next section be taken as a point of departure to identify "knots and nodes" in which nature and gender relations intersect.

All Living Beings Interlinked in a Complex System

In *Bird People*, as well as traditional images and sounds of the airport as a site of high technology, natural spaces are also present in the form of plants, animals and environments. During her explorations as a sparrow, Audrey comes to an area of woodland where homeless people sleep in their cars at night. This scene shows the precarious existences of the airport workers, including Audrey and Simon the night porter, who (not just metaphorically) enter the hotel by the back door. At night, Audrey-as-a-sparrow is chased by an owl, and slips inside a car through a window that has been left cracked open. What looks at first glance like a pile of bags turns out to be a person sleeping in the car, covered in blankets. Viewed through the lens of nature relations, brief moments that seem tangential to the film's main narrative also assume significance. A fellow passenger on the airport train ponders whether he should cook a whole rabbit or just half; Audrey watches a sparrow that peers in through the train window; during a Skype call with Gary's wife Elisabeth, a sparrow appears at the kitchen window; memories of fishing are shown in a flashback and then mentioned in conversation; an elephant-print shirt worn by Gary's colleague; a proverb about rabbits and chickens; a cat outside the hotel entrance and later in the hotel corridor; painted sparrows; crows, an owl; a flock of birds. By contrast with the organized territory of the airport, the car park in the woods is an indeterminate, marginal space, what Clément (2010) terms a "third landscape". This third landscape is at once an abandoned plot of land and a sanctuary of biodiversity.

Viewed through an ecological lens, the natural space on the periphery assumes greater significance. Recent work in the environmental humanities (including Bauer, Güttler and Schlünder's 2019 study on Frankfurt Airport) has shown how, as global hubs and sites of digitalization and high technology, the world's airports are part of ecological systems. If, following Gilles Clément (2015), we understand ecology as "an observation through which all living beings find themselves bound in a complex system including humanity, air, water, rocks and invisible energy fields", then ecology includes urban wastelands, motorway verges and the natural spaces of airports. These spaces gain in significance if the airport is viewed through an ecological lens. It is practically impossible to appreciate this significance outside the realm of cinema, due to the strict security regimes at airports. Passengers go about their routines on autopilot: those with hand luggage only proceed directly to check-in; if there is time, they pop to a restaurant or do a little shopping; then they make sure to remember to go to the toilet and line up for boarding. It's only if there are unexpectedly long waits or people find themselves stranded that the airport's ecological nature is temporarily laid bare.

One such moment occurred in July 2017 as I was returning from the NECS (European Network for Cinema and Media Studies) annual conference. After an unknown person went the wrong way through a door in the security zone, Charles de Gaulle Airport was brought to a standstill for several hours. Aimlessly wandering round the airport like Gary and Audrey, I discovered a sheep pen behind the airport terminal's huge glass facade. It felt like a Christian religious phenomenon. In that moment, it was impossible to find any explanation for why it was there; a brief Google search came up with nothing. Only later on, after more thorough research, did I find a post on an airport chat forum[6] that confirmed the sheep really had been there.

The sight of a sheep pen in the middle of one of Europe's biggest airports felt like something from a bygone era. Its magic came from difference; it can be interpreted as a material–semiotic node or knot, where differences such as nature/culture and nature/technology intersect. Audrey's transformation into a sparrow possesses similar qualities. In the flight of a bird, the logistics of an airport are perceivable from another perspective. This ecological perspective on the airport is initiated by a Harawayian "figure" that interweaves gender, anthropological and class difference in such a way that they become indistinguishable. The fairy tale-like conceit of Audrey's transformation into a sparrow opens up an alternative perspective on the airport that makes natural spaces visible and causes its role as a complex logistics hub to recede into the background. *Bird People* thereby creates an experiential space in which the idea of the airport as part of an ecosystem is vividly palpable. The bird's perspective – which, mediated through the character of Audrey, we are able to fictively adopt as spectators – reveals the airport in a different light. Until that point, the film's world has been defined by flexible people in non-places; Audrey's transformation marks a fundamental shift in perspective and in what the film picks out as relevant. Things that are peripheral, precarious or invisible to the human eye are brought to light: a bin, the loading ramp at the back of the Hilton, stories of precarious existences and, above all, animals and plants. When Audrey becomes an animal, the airport is revealed as a site of diverse potential and intersecting differences.

Summary

Beginning with the book object *The World We Live In*, which was produced by Hörner/Antlfinger's interspecies collective CMUK, I examined two concepts coined by Haraway and Derrida that have been highly influential in gender studies and cultural animal studies respectively. This was followed by a case study of the film *Bird People*, which demonstrated how anthropological, gender and class difference intersect and can open up new perspectives on a place. In the third section, I turned my attention to Charles de Gaulle Airport itself, and situated it as part of a totality of living beings within a complex ecological system. I interpreted the anachronistic-seeming sheep pen at the airport as a material–semiotic knot or node that – like Audrey's flight (as a sparrow) around the airport – is able to open a different perspective on the place. From this perspective, the airport is revealed as a space where differences such as nature/culture and nature/technology become intertwined. In *Bird People*, this ecological perspective on the airport is initiated by a transition from individual gender, anthropological and class differences into a proliferation of difference. In this proliferation of difference, we can discern the historicity of the idea of individual difference.

Translated from German into English by Andrew Godfrey.

Notes

1 *Die Welt in der wir leben*, 2016. Book object, display case 50 × 50 × 170 cm, photographs 40 × 60 cm. https://h--a.org/de/cmuk/de-die-welt-in-der-wir-leben/
2 On the distinction between human animal studies and cultural animal studies, see Borgards (2012).
3 According to Haraway, this episode is primarily an excuse for Derrida to talk about philosophy. See Haraway (2008, 20).
4 On the debate about the exchange of human and animal gazes that was sparked in media cultural animal studies by John Berger's essay "Why look at animals?" (2009), see Pick (2015) and Burt (2005).
5 Pascale Ferran in conversation with Birgit Kohler on 3 November 2017 at „Werkschau Pascale Ferran" at Arsenal – Institut für Film und Videokunst, Berlin.
6 "Yesterday walking from Terminal 2F to the train station I spotted a bunch of sheep – I hadn't noticed them before. Anyone know why they are there? Cheese for inflight catering? ☺" (Sea-Flyer 2016)

References

Augé, Marc. 1992. *Non-Places: Introduction to an Anthropology of Supermodernity.* Translated and edited by John Howe. London and New York: Verso.

Bauer, Susanne, Nils Güttler and Martina Schlünder. 2019. "Encounters in Borderlands: Borderlining Animals and Technology at Frankfurt Airport." *Environmental Humanities* 11 (2): 247–279. https://doi.org/10.1215/22011919-7754445

Berger, John. 2009. "Why Look at Animals?" Chap. 1 of *About Looking.* London: Bloomsbury.

Borgards, Roland. 2012. "Tierphilosophie, Tiertheorie und Teleosemantische Differenz." *EWE* 23 (1): 41–44.

Burt, Jonathan. 2005. "John Berger's 'Why Look at Animals?': A Close Reading." *Worldviews: Environment, Culture, Religion* 9 (2): 203–218. https://doi.org/10.1163/1568535054615321

Chakrabarty, Dipesh. 2017. "The Politics of Climate Change Is More Than the Politics of Capitalism." *Theory, Culture & Society* 34 (2–3): 25–37.

Clément, Gilles. N.d. (2003). "The Third Landscape." http://www.gillesclement.com/art-454-tit-The-Third-Landscape

Clément, Gilles. 2015. "'Garden' Defies Cultural Divisions." *La lettre du Collège de France* 7. https://doi.org/10.4000/lettre-cdf.2639

Derrida, Jacques. 2008. *The Animal That Therefore I Am.* Translated and edited by David Wills. New York: Fordham University Press.

Ferran, Pascale, dir. 2010. *Bird People.* Archipel 35, DVD.

Förster, Annette. 2012. "Actresses Interacting with Animals in the Silent Cinema: Kathlyn Williams, Berthe Dagmar, Musidora, Nell Shipman." In *Animals and the Cinema: Classifications, Cinephilias, Philosophies,* edited by Sabine Nessel and Winfried Pauleit, 66–86. Berlin: Bertz & Fischer.

Haraway, Donna J. 2008. *When Species Meet.* Minneapolis: University of Minnesota Press.

Haraway, Donna J. 2016a (1985). "A Cyborg Manifesto: Science, Technology, and Socialist-Feminism in the Late Twentieth Century." In *Manifestly Haraway,* edited by Donna J. Haraway and Cary Wolfe, 3–90. Minneapolis: University of Minnesota Press.

Haraway, Donna J. 2016b. *Staying with the Trouble: Making Kin in the Chthulucene.* Durham, NC: Duke University Press.

Meyer, Katrin. 2017. *Theorien der Intersektionalität zur Einführung.* Hamburg: Junius.

Nessel, Sabine. 2019. "Vom Gorilla zum Showgirl: Der Diskurs des Anderen in Josef von Sternbergs Film Blonde Venus." In *Im Verwandeln der Zeit: Reflexionen über filmische Bilder*, edited by Regina Brückner, Bernhard Groß, Matthias Grotkopp and Eileen Rositzka, 121–130. Berlin: Vorwerk 8.

Ngai, Sianne. 2006. "Black Venus, Blonde Venus." In *Bad Modernisms*, edited by Douglas Mao and Rebecca L. Walkowitz, 145–178. New York: Duke University Press.

Pick, Anat. 2015. "Why Not Look at Animals?" *NECSUS* 4 (1): 107–125. https://doi.org/10.5117/NECSUS2015.1.PICK

Sea-Flyer. 2016. "Sheep at CDG?" Post on the Flyer Talk Forum. 2 July 2016, 1.02 a.m. Accessed 6 August 2021. https://www.flyertalk.com/forum/air-france-frequence-plus/1775702-sheep-cdg.html

Sennett, Richard. 1998. *The Corrosion of Character: The Personal Consequences of Work in the New Capitalism.* New York: Norton.

Ullrich, Jessica. 2016. "Schwarmästhetik: Distributive Agency in der Interspezies-Kollaboration." In *Die Welt in der wir leben: Hörner und Antlfinger und das Interspezies-Kollektiv CMUK: Ausstellungskatalog.* Cologne: Academy of Media Arts Cologne.

Wiazemsky, Anne. 2009 (2007). *Jeune fille.* Munich: Beck.

Wild, Markus. 2008. *Tierphilosophie zur Einführung.* Hamburg: Junius.

8 Wildlife Among Us. Post-Natural Worlds and Interspecies Encounters in Nicolette Krebitz's *Wild*

Andrea Seier

"Now I'm gonna fix us some breakfast" opens one of only a couple of cheerful, brightly lit scenes in Nicolette Krebitz's 2016 film *Wild*. It takes place between protagonist Ania and her new flatmate, a wolf, and is preceded by a – visually and musically – spectacular masturbation scene in which bannisters play a key role, and in which the wolf has its share in some ways. Ania's new flatmate did not exactly choose to move in with her. Instead, Ania enlisted the help of migrant workers from Southeast Asia in a second-hand clothing factory to catch and temporarily anaesthetize the animal. They do this by using a so-called fladry line, which Ania researches in advance via a Google search. This filmic device simultaneously informs the viewer about this hunting method used to capture wolves for research and relocation purposes. Ania, however, has a different purpose in mind, although the term "relocation" does indeed fit her carefully mapped out and resolutely executed plan. She is also engaged in a kind of research, albeit neither of the zoological kind nor linked to an academic discipline. Ania's research rather falls within the category of what Felix Guattari (2000, 44) calls the "logic of intensities" in his text *The Three Ecologies*, or what he describes as "social ecology" (41). That is, it raises existential questions.[1] As is apparent, Ania does not pursue these fundamental questions with energetic rigour. In a way, she stumbles upon them through her encounter with the wolf, which is when these issues suddenly become unavoidable.

The film *Wild* discusses existential questions to which the answers are unknown, yet not entirely hopeless. What does become clear is what kinds of relationships are *no longer* possible or should not exist anymore: living in couples, work relationships, relationships with our surroundings and the environment. *Wild* – through Ania's more or less chance encounter with the wolf – forces a revision and readjustment of all of the above. Tanja Prokic states aptly that the film *Wild* is concerned with asking questions which are beyond answerable: "how not to be governed in a neoliberal world?" (2017, 134).[2] The central suggestion here is hence not the famous *walk on the wild side* Lou Reed recommended in his song, advocating an experimental approach to compulsive heterosexuality, first of all.[3] The film is about the domestication of a wolf, which happens in parallel with protagonist Ania's

DOI: 10.4324/9781003023319-12

Figure 8.1 Film poster (*Wild*, dir. Nicolette Krebitz, 2016).

progressive and gradually increasing wildness. Both aspects – domestication and feralization – form the film's lines of flight and signify a search for new assemblages, which are rather distant from her life in a prefabricated high-rise building, her job in IT and some heterosexual romances (Figure 8.1).

The film connects Ania's newly found wildness with the question of the (im)possibility of female subjectification and, as such, clearly follows in the footsteps of its feminist predecessors (see Showalter 1981).[4] I argue that the film manages to assert the attractiveness of the wild as an instance of resistance while eschewing the traditional binary of nature versus culture. Rather, the topos of the female feralization is updated for our time, taking into account the conditions of a post-natural world.[5] In Nicolette Krebitz's film – similar to Donna Haraway and/or Bruno Latour – wilderness, traditionally determined by the absence of human agency, paves the way into an environment which is socially permeated by multiple human and non-human species. Their encounter is neither coincidental nor fleeting, but rather the result of their living conditions. They open up to each other, become responsible for each other and strive towards a (or multiple) shared future/s.

In the following section, I will begin by reconstructing several general theories about the relationship between wilderness and femininity, including the most recent developments in feminist theory. Subsequently, I will engage with the question of how the film *Wild* updates the topos of a female becoming wild. In this, I am particularly concerned with the question of whether the film manages to conceptualize the wilderness as part of a post-natural world.

Interpellations from This World and Beyond: A Female Feralization

Wilderness is a utopia which – as a feminist topos – is not new. It runs through feminist theory as it does through literary and film history.[6] It more or less consistently updates the search for the margins of civilization, for the breaks and blind spots in a patriarchal culture, in a way that will outlive even post-feminism, as Krebitz's film suggests. Although its genealogies date back much further, the topos of the wild feminine "benefitted" in the twentieth century not least from Freud's ([1931] 1963) famous association of the feminine with a dark and unknown continent. In feminist theory, this becomes a notion of the feminine as something not recognized by patriarchy and therefore not domesticated, which turns into the idea of a separate form of resistance, developing more or less under the radar (see Griffin 1978, as one example among others). In essentialist versions of female wildness, the feminine is portrayed as the opposite of civilization with its masculine connotations, although the borders between essentialist and deconstructionist versions are much more blurred than they appear at first sight.[7] This form of ambiguity is also tied to another: the notion of the wild feminine oscillates between utopia and criticism, between reality and fiction.

The idea of the feminine Other may be outdated now. But that does not mean that the opening up of a *space-off (either* in fiction, theory or in "real" life), where existing requirements of successful femininity or all other gender performativities may be rejected or at least scrutinized, is over and done with (see De Latis 1987). The notion of the female becoming wild, as Krebitz's film demonstrates, is still promising and worth telling. It highlights consequences for women's environments, both on film and in real life.

One of the main differences between Krebitz's film and its predecessors, such as Marlen Haushofer's ([1963] 2013) novel *The Wall*,[8] is that the situation of woman and wolf cohabiting in a small flat is freely chosen and not forced upon the protagonist by adverse, catastrophic circumstances. In addition, the encounter between female employee and wolf appears much more plausible because of the film's setting in Halle (Saale) in Saxony-Anhalt.[9] The reference to an existing place moves the encounter between Ania and the wolf into the realm of the possible instead of situating it within utopian or fantastical spheres. This point is worth noting not least since the topos of the wild feminine previously received its critical feminist dimension through conjuring up a utopian possible world, an alternative to a male-dominated civilization.

In 2015, a single she-wolf settled permanently at the Ohrdruf military training ground not far from Halle. According to the German nature conservation society Naturschutzbund Deutschland e.V. (NABU), due to "a lack of male wolves" the she-wolf mated a little while later (ca. 2017) with a dog and gave birth to a litter of wolfdog puppies. For "reasons of conservation", the NABU website states, "shooting permits had been issued for the puppies". Since 2019, the she-wolf has paired with a male and has thereby laid the foundations for establishing the "first Thuringian wolf pack".

This Thuringian wolf pack, living an urban life on a former military training ground, must be regarded as providing inspiration for Krebitz; albeit it is not directly referenced in the film.[10]

Nature, here, is not portrayed as the simple opposite of culture, nor does the film posit a particular affinity between women and the natural world,[11] as Susan Griffin (1978) does in her eco-feminist classic *Woman and Nature*. Griffin's subtitle, "The Roaring Inside Her", however, is still appropriate for Krebitz's film. In the last third of the movie, it is directed at everything that gets in Ania's way.

Against the backdrop of current gender and nature relations, the question arises what kind of promise (or unlearning) the film aims to conjure up by invoking the wild? How are femininity, nature, culture and wildness related to each other in the film? Does it open up a space for an understanding of gender in a more-than-human sense? In other words: is the wild here connected with emancipatory thought that moves away from an anthropocentric point of view?

Feminist New Materialism

Feminist new materialism is regarded as one of the more heterogeneous research areas that have been gathered together under the umbrella term the "nonhuman turn", alongside Animal Studies and Object-Oriented Ontologies (see Grusin 2015). What these research areas have in common is their concern with rethinking "human" subjectivity while paying closer attention to dynamic relationships with the non-human and the inanimate world. The category "gender" is hereby related to (individual) bodies, but mainly serves to communicate the insight that boundaries between bodies are drawn by various interconnected discursive and material relations which guarantee or prevent the bodies' viability. The establishment of boundaries with discourse theory within feminist new materialism mainly arises from rejecting the notion of a passive matter which only gains meaning through discursive practices. This rejection leads to various conceptualizations of discourse and matter, politics and nature, ranging from alternative ontologies (e.g. Barad 2012) to feminist speculations (e.g. Angerer and Gramlich 2020) and an update on vitalist positions (e.g. Bennett 2011). Even though these approaches may conceptualize the relationship between nature, gender and politics differently, they all pursue the same goal of challenging social-constructivist short-cuts within Gender Studies. In other words: the question about how a compulsory heterosexual order may materialize in individual bodies is currently supplemented by a discussion around what kind of various materialities make up the category "gender". Paul Preciado's (2008) self-experiments with testosterone and his diary entries marked by theoretical engagement have provided an "illustrative" contribution to the debate. Preciado shows how many specialized discourses (e.g. biology or medicine), materials and organs (e.g. bones, muscles, hormones, medical and psychotropic drugs) and practices (e.g. writing, eating, sleeping, sex) meet within one's supposedly "own" body. After reading this personal book, the notion of one's "own" or "human" body seems like a useless abstraction. Instead, one's focus is directed towards the wide range of interferences between human and non-human agency.

The extent to which these interferences make up the conditions for our being in the world is Karen Barad's research focus. The physicist has developed her own stance based on a combination of quantum physics, Queer, and Performativity Studies. Using atoms, foetuses, frogs, and lightning, Barad (2011) advocates for "Nature's Queer Performativity" which stretches the concept of queerness to the max. Studying queer identities and practices – that is those which eschew unambiguity – "only" through human life-forms constitutes an oversimplification, Barad argues. Instead, research must bear in mind that both historical and present-day queer practices have given rise to questions about the boundaries between human and non-human behaviour. This is the case, for example, when queer practices have been labelled as "acts against nature" and "bestiality" (121). On the one hand, humans are accused of these acts "against" nature, a concept which implies that nature must be protected from human acts. At the same time, the actors are accused of "bestiality", that is, of acting like animals, that is in a non-human manner and as part of nature. The accusation of human bestiality therefore trips over the very separation between nature and culture it is trying to safeguard:

> What kinds of acts against nature inspire moral outrage? Queer pleasures for sure, even some forms of heterosexual sex, and an assortment of other human practices. Clearly, the nature/culture divide is at issue and at stake, but the logic that tries to hold it in place is quite perverse. (121)

Against this backdrop, Barad becomes – as Donna Haraway did before – interested in the term "critter", which allows for the divide between the human and the non-human to be bracketed. Critters, according to Barad, "do not have inherently determinate identities, by definition" (126). In other words: critters do not so much follow an identity-based but a *queer* logic since they conjure up – simultaneously and also literally – a relationship with and demarcation from the human.

Looking at the proximity between woman and nature in the context of essentialist feminism, an interesting variation emerges from the current form of feminist theory. One the one hand, the focus shifts onto historical dis/continuities in the constitution of the non-human other, which include tactics of dehumanization in favour of nature. On the other hand, the serious consequences of a denial of interferences between nature and culture come into view – not just, but *also* in the sense of shaping, transforming and exploiting nature.[12]

In this setting, gender differences are implied rather than explicitly addressed. That is why new materialism often needs to meet the objection that it not only problematizes anthropocentric thought but also reiterates it in its repeated reference to the universal "we" (see Lettow in this volume).[13] Another subject of ongoing discussion is where and how an intra-active ontology, as posited by Barad, can gain insights and experiences. However, through stressing the extent to which human action, thought and feelings have always been embedded in their respective environments, Barad and authors such as Donna Haraway (2016) and Anna Tsing (2015) have provided self-reflexive insights and a new impetus for ecological Gender Studies (see also Bath et al. 2013).

At the same time, studies within new materialism sometimes struggle with the distinction between justified objections to previous schools of thought, and insistent rhetorics of the new (see Seier 2014). However, this distinction is of no great importance. What is more important is the (age-old) debate about an understanding of materiality which allows us to think and, wherever necessary, change the various forms of social, political and discursive situatedness of materiality and bodies. This question is far from being a new one. Yet, it must be re-issued again and again to save critical objections to existing and fictitious notions of future relationships between nature and gender from becoming null and void (see in this context also Meißner 2014).

Instead of stressing the *new* within new materialism, it appears to be more urgent to work on an update on the intersectional conditions of forms and practices of being in the world. Against the backdrop of current trends towards optimizing life, the body, and work, which dramatically affect and change all of the above, this undertaking seems all the more necessary. It could also help to discuss the shared foundations of poststructuralism, historical and new materialism[14] as well as possible intersections with related disciplines such as Postcolonial and/or Disability Studies (see Habermann 2017; Schadler 2017).

However, Brigitte Bargetz found out the opposite. She discusses the disintegration of "new" and "old" materialisms, grounded in the paradoxical term "materiality" itself, which new materialism, in her view, understands simultaneously in too broad and too narrow a sense:

> The term proves too narrow since it relates primarily to matter, objects, artefacts, materials and the non-human, while neglecting the materiality of social relations and not least gender relations. In combination with a concept of materiality that is too broad, and which insists on not tying society to the limits of humanity, this proves politically problematic.
>
> (Bargetz 2017, 139; transl. DeWald)

Bargetz calls this trend "writing out the social"[15] within new materialism. With Bruno Latour's distinction in mind between a traditional understanding of the social as its "own" sphere (sociology of the social) and a concept of the social which is predominantly interested in emerging associations and assemblages (sociology of associations), such as in Actor–Network Theory,[16] Bargetz's argument appears as a legitimate demand: instead of excluding each other, both approaches should benefit from each other in order to strengthen intersectional perspectives on materiality.

Even though Bargetz's view from the "outside" only partly does justice to these heterogeneous new materialist approaches, she simultaneously draws attention to a division of labour within Gender Studies which is at least worthy of discussion. She shows that, so far, neo-materialistic approaches have only rarely been combined with studies whose focus lies elsewhere, such as Governmentality, Affect, or Postcolonial Studies.[17] Problems like these also resonate through current forms of protest in which ecological and social perspectives only slowly begin to intertwine. Nicolette Krebitz's film *Wild*, however, manages to combine some of these issues which are generally being discussed separately – within Gender Studies, and also in public. I will discuss this assumption more closely in the last section of my chapter.

New Materialisms and Old Problems in *Wild*

The film *Wild* focusses on two "critters" which are labelled as human and animal in a way Barad might find too restrictive. Yet, it is their ever closer encounter which drives the film's narrative ("plot" is not the appropriate term here). "Acting" only happens across creature boundaries and includes setbacks and mutual hindrances.

Here, the wolf is not or at least not *just* a metaphor or the product of cultural imagination. In the sense of new materialism and its critique of representationalism,[18] the wolf is a physical opposite within the urban reality of the film. This posed a challenge when filming since director Krebitz refrained entirely from using trick shots and visual effects (see an interview with Nicolette Krebitz at *laserhotline* 2016). The wolf's presence in the city appears neither in a romantic nor in a catastrophic light. Yet, for protagonist Ania, it triggers a decisive transformation pushing her both to the limits of what it means to be human, and especially the more restrictive borders of femininity.

At the beginning of the film, Ania seems introverted, shy, determined by external forces rather than herself, without ties to places or people, with the exception of her dying grandfather. Insofar as the film refuses to engage in deep psychological diagnoses of its characters or to offer a dramatic narrative, the spectator – in the literal sense of the word – simply watches Ania. The wolf as the object of desire comes as a surprise. What is unexpected as well is that Ania lets herself be touched by it at all – through a single glance on her way home from work. When looking at her colleagues and others, Ania usually stares into the void. Yet this encounter is characterized by their eyes meeting. And these reciprocal glances create space for projections, utopias, affects and desire, however direct or diffuse these may be. The development, in any case, occurs in multiple directions, and it is rather shown in and through images than narrated. Ania's gradual process of becoming wild goes hand in hand with astonishing camera effects by cinematographer Reinhold Vorschneider, including surprisingly radical and physical acts and excrements of all kind transcending the boundaries of the body.

Ania's encounter with the wolf sets herself in motion. It causes an affective transformation which triggers her own actions – a drive she has lacked up until now. This transformation process propels Ania forward and also guides the spectator through the film. Although it is not a linear development, once set in motion, the process is irreversible.

The film is clever enough to take affects seriously, without limiting them to the story of stunted or unrequited female desire (although a degree of this is present throughout the film).[19] Instead, manifest and diffuse latent intentions, desire and affect alternate in these relationships marked by appropriation and exploitation, power and discursive disposition, without fully developing as any of them. The film is interested in the spaces between the manifest and the latent, between affect and desire. This is where it situates the "wild". Wildness, therefore, is not a synonym for the absence of human agency, but rather "happens" in the fissures of discursive power. Requirements of femininity which can extend to an expected subjugation in the workplace are one example of this. All of these notions are not told but shown, such as in a scene where Ania's boss makes her bring coffee to his office. Ania's relationship with the wolf, however, who moves into her sister's room, is not frictionless either. Ania observes him through a hole in the wall, yet shortly after the wolf tears down the wall, ending the surveillance which had turned the animal into an object. This event impacts the spectator's gaze as well: the film is interrupted for a brief moment. The wolf on screen, the effect suggests, is challenging the fourth wall between animal and audience.

What makes the portrayal of this cross-species flatshare so special is that it is shown to be unusual, yet never dramatic. The wild and the everyday are situated in close proximity to each other. The staging of their shared life is too realistic to be dismissed as Ania's phantasy alone: she still needs to go to work and organize daily life in their shared flat, including mealtimes.[20] Yet, the film also exploits its aesthetic possibilities through a highly reflective use of vivid imagery to push the boundaries between the everyday and the utopian. The film is (thankfully) not content with realism alone.

Once adverse circumstances have been overcome and the new flatmate has successfully moved in, a new life can begin in the shared flat. This requires compromises on both sides: DIY protective clothing made from outdoor gear on the one, and curtailed freedom of movement on the other side (Figure 8.2).

Figure 8.2 Ania and the wolf in the flat (*Wild*, dir. Nicolette Krebitz, 2016).

Those who long for nature also need outdoor equipment to protect themselves from it. Ania, however, needs these protective items (such as a helmet and a sleeping bag) in her own home. And the evocative scene in which Ania feeds the wolf succeeds not only on an aesthetic level, but it also pokes fun at her pragmatic solutions to outdoor problems in indoor spaces. As such, it reiterates a decidedly feminist topic.

The shared life of woman and wolf in a prefabricated high-rise building[21] makes the supposedly special affinity between woman and nature recede into a dim distance. What began as the domestication of the wolf, gradually turns into Ania's growing ever more wild. And her desire for the wolf – as *one* of the effects of their encounter – turns into Ania's becoming-animal. Using Haraway's (2008) terms, in meeting the wolf, Ania encounters her significant other.[22] Her "turn" to the wild is the opposite of a reconnection with her self, but is rather a departure for something (new) which could even exceed the "self". This development calls for a positive turn towards what Haraway describes as an internal exteriority or what Trinh T. Minh-ha (2010) calls "[e]lsewhere within here". Once set in motion, the question of the "right life within the wrong one" becomes urgent for Ania, without getting caught up in a dialectical closure. In his review of the film, Ekkehard Knörer (2016) calls this development a "becoming wild as self-technology". He writes:

> The wolf is in the forest, but because the forest is in the city, this wolf is not simply wild but has already got lost and lost itself in civilization. Ania domesticates it a little, while becoming pretty wild herself. The wolf becomes her means to an end, a tool used to re-wild her self: becoming-animal takes the place of IT. Fellow humans become wolf.
>
> (Translated by Rebecca DeWald)

Krebitz pulls a lot of punches in the staging of this transformation: in contrast with Marlen Haushofer's protagonist, who associates the act of becoming non-human with the abyss, in Krebitz's film the non-human is connected with liberation. The breakfast scene mentioned earlier shows Ania, frying eggs, in a one-sided "couple's dialogue" in which she explains "human" eating habits to the wolf, the very same habits to which she is about to bid farewell. "Do you think I'm fat?", she initially asks. The question, which involuntarily pops into her head, appears to her like an echo from bygone times when women still cared about their appearances. The wolf does not care about Ania's weight,[23] so she serves coke for breakfast – a drink that used to be banned from her diet – while her smartphone, showing a call from her office on the display, has already lost much of its appeal. And there will be more changes than coke for breakfast, triggered by her act of going wild. Not least further sexual encounters with her boss and cleaning staff in the office, featuring blood, shit and sperm. Her boss, played by Georg Friedrich, who would pace up and down in his glazed-off office at the start of the film, as if in a cage, later observes Ania's becoming-animal with irritation bordering on envy. "I will build you a stable for your dog!" he will call out to her, both petty and aggressive at the same time, later in the film when she uses the wolf to defend herself against him. Gender relations are more than an endured domestication here. The spectator also notices the boss's radical underestimation of the situation, which may be extended to an underestimation of femininity in the symbolic order of things and symbols Everyone (her boss, colleagues, the pet shop assistant and even her own sister) underestimate Ania in the beginning. The only exception is her beloved grandfather, who dies when she starts becoming animal. When Ania finally, as a form of goodbye, shits on her boss's desk, her underestimation comes to an unmistakeable end: the "dark continent" strikes back. These fulminant scenes blur the lines between the everyday and utopia, between pragmatism and reverie. Rather, these concepts depend on each other to envisage a form of survival that is more than just coping.

The film suggests, without employing further sociologically founded analyses, that its other female characters are also confronted with the question of female subjectification, which, in brief, is impossible without a "becoming-inhuman". Insofar as the film closely links humanity with the unreasonable demands of gender expectations, the rejection or, at least, pleasurable relativization of the human gives it a feminist baseline.

Hybrid Futures: Neither Natural nor Social

Eroded boundaries between nature and culture are both the starting point and the line of flight of this cinematic narrative. Animals in cities (at the beginning of the film) and women in the landscapes (at the end of the film) are neither good nor bad. While the strict demarcation between nature and culture has led to the destruction and exploitation of the natural world, the irritation of this boundary also offers prospects of a departure. This departure must not be misunderstood as self-fulfilment, nor as necessarily feminine. If anything, the film suggests towards the end a utopian genesis of a companion species in Haraway's sense, and regards the relevance of gender binarisms to be at least questionable. The type of relationship possible between woman and wolf still needs to be learnt or invented. It oscillates between friendship, love and a community of solidarity, rendering strict borders between them defunct.

The film's final sequence shows Ania and the wolf walking through the *naturalcultural* desert of the Lausitz brown coal field,[24] sometimes together, sometimes separately, at times sharing a drink from a puddle (Figure 8.3).

Not much grows in the Lausitz "natureculture" nowadays. Yet, Ania flourishes: a companion species to come. Her flourishing is not a natural development capable of leaving civilization behind. Rather, it is situated in a highly competitive social, existential terrain, which is currently contested in the sense of Guattari's three ecologies (the environment, social relations

Figure 8.3 Companion species in the Lausitz brown coal field (*Wild*, dir. Nicolette Krebitz, 2016).

and human subjectivity) mentioned above: the Lausitz brown coal field is more than a visual framework for pressing current eco-social questions about phasing out fossil fuels and climate protection. It is much more than just a backdrop. It rather represents a battleground within the struggle for the terrestrial conditions of existence, which also implies the power struggle over who gets to shape the future.[25]

The area in which the film ends will be difficult to inhabit in the long run. The woman embarking on her becoming-animal has hence no home (yet) – although this does not mean that she left a home behind, as Knörer (2016) confirms. After all, there were reasons for her change of course. Leaving civilization behind, however, is not an option. The myth of the "lone wolf" is not appealing either and has already been taken anyway. What remains is the rejection of an identitarian and exploitative logic, which offers a ray of light at the end: laughter and sunshine take the place of IT. "How not to be governed in a neoliberal world?" (Prokic 2017, 134). Although the film does not answer this question, it makes a start.

Translation from German into English by Rebecca DeWald.

Notes

1 Guattari assigns these questions to three registers, which he calls "ecological" because of their interconnectedness (within themselves and among each other): the environment, social relations and human subjectivity (28). The relevant text passage reads: "Social ecosophy will consist in developing specific practices that will modify and reinvent the ways in which we live as couples or in the family, in an urban context or at work, etc. Obviously, it would be inconceivable to try and go back to the old formulas, which relate to periods when the planet was

far less densely populated and when social relations were much stronger than they are today. But it will be a question of literally reconstructing the modalities of "group-being" [*l'être-en-groupe*], not only through "communicational" interventions but through existential mutations driven by the motor of subjectivity. Instead of clinging to general recommendations we would be implementing effective practices of experimentation, as much on a microsocial level as on a larger institutional scale" (Guattari 2000, 34–35).

2 This question picks up on Michel Foucault's (1996) understanding of criticism in "What Is Critique?".

3 The process of Ania's becoming wild has already set in when her boss tries to reassure her by saying: "You'll see, a couple of weeks and you'll be your old self again". To which she replies: "But I don't want to be the way I was".

4 See the following section.

5 For an introduction to the debate about nature shaped and controlled by humans, see, for example, Horn and Bergthaler (2019).

6 Marlen Haushofer's novel *The Wall* (1963) is one of the most prominent examples here (see also, e.g. Hall 2016, Lawrence [1923] 2006 and Varda 1985).

7 Marlen Haushofer's novel *The Wall* is also a very good example of this ambiguity. On the one hand, femininity is contrasted with masculine civilization. However, the concept of femininity itself loses its meaning in the course of the novel.

8 The novel was made into a film in 2012, directed by Julian Roman Pölsler.

9 However, there is no need for the audience to identify the city in the film as Halle.

10 In interviews, Nicolette Krebitz cites the story as *one* of her sources of inspiration among many.

11 In contrast, Pölsler's film *The Wall* is set in an Alpine valley providing a far too perfect backdrop for actor Martina Gedeck's performance. In this setting, nature follows the course of the seasons, untouched by climate change. And when the protagonist and her dog move up to the alp for the summer, the mountain pasture is flooded with golden light and the faces of woman and dog are so at ease that parallels with the sentimental *Heimatfilm* aesthetics are distressingly obvious. On the occasion of the film version, Stefan Herbrechter reread the novel through a posthumanist lens and argues for a "'proto-posthumanist' affinity between Haushofer's novel and the 21st century present day" (2014, 48; transl. DeWald). Herbrechter focusses on the subject of writing within the novel, in which he traces a movement from *écriture féminine* towards *écriture animale*.

12 Of course, this was also the goal of earlier approaches to ecofeminism. The film picks up on both schools of thought, while greatly enjoying various kinds of border-crossing.

13 The turn towards ontology leads to the same critical engagement.

14 The relations between historical and new materialism are discussed in Lettow (2018).

15 Bargetz also cites counterexamples, albeit the most prominent name – Donna Haraway – is missing from her list. In contrast, Bargetz positively highlights Mel Y. Chen's study on the US debate about the intoxication of the middle classes through toys produced in China. In this study, Chen (2012) discusses the way in which inanimate matter (such as toys) is somewhat "animated" through global capitalist mechanisms of power and racism.

16 The sociology of the social therefore designates the "social forces at work" "'behind' the actors' backs" (Latour 2005, 136) which forms a sphere encompassing its own rules and laws. The social as association, however, does not form its own sphere, nor is it bound by (invisible) structures. It only becomes visible in the traces left behind by forming new associations. Following Actor–Network-Theory, the social stands for "an association between entities which are in no way recognizable as being social in the ordinary manner" (65).

17 This does not apply to the book by Bath et al. (2013).

18 Representational concepts which rely on optical metaphors, such as reflection, mimesis or similarity, are generally in opposition to processes of becoming, which Barad advocates. This is because they stress the separation between observer and object, instead of addressing their reciprocal emergence (see Barad 2007, 86–87).

19 References to Nagisa Oshima's (1986) film *Max, Mon Amour*, which discusses female desire through a woman's (Charlotte Rampling) relationship with an ape, are recognizable, yet do not impose themselves.

20 On her way back from work, Ania asks at the supermarket meat counter what to get for a large dog "if it's a special occasion"? Rituals of romantic relationships between two people are simultaneously exercised and displaced.

21 I refer here to the uniform housing estates built in East Germany before the fall of the Berlin Wall using a large panel system (*Plattenbau*). However, these were called "new builds" (*Neubauten*) before the German reunification.

22 By employing the concept of companion species, Haraway seeks a way out of the anthropomorphization of the animal world, on the one hand, and the erection of absolute boundaries between animal and human, on the other. In its place, Haraway puts a turn towards a partial connection and significant otherness, for which the relationship between humans and dogs forms only a starting point.

23 Unfortunately, the film is not entirely indifferent to its protagonist's body shape. At times, it employs fragmented images of Lilith Stangenberg's body in a way which Laura Mulvey criticized as patriarchal visual representation.

24 Much like the city of Halle, it is not necessary for the spectator to recognize the location. The visual impression of this natural-cultural landscape is of importance here.

25 See here, for example, actions by environmental activists, such as *Ende Gelände* in the Lausitz region.

References

Angerer, Marie-Luise and Naomi Gramlich. 2020. *Feministisches Spekulieren. Genealogien, Narrationen, Zeitlichkeiten*. Berlin: Kadmos.

Barad, Karen. 2007. *Meeting the Universe Halfway. Quantum Physics and the Entanglement of Matter and Meaning*. Durham, NC: Duke University Press.

Barad, Karen. 2011. "Nature's Queer Performativity." *Qui Parle* 19 (2): 121–58. https://doi.org/10.5250/quiparle.19.2.0121.

Bargetz, Brigitte. 2017. "Writing Out 'the Social'? Feministische Materialismen im Streitgespräch." In *Material turn: Feministische Perspektiven auf Materialität und Materialismus*, edited by Christine Löw, Katharina Volk, Imke Leicht, and Nadja Meisterhands, 133–150. Opladen: Barbara Budrich.

Bath, Corinna, Hanna Meißner, Stephan Trinkaus, and Susanne Völker, eds. 2013. *Geschlechterinterferenzen. Wissensformen-Subjektivierungsweisen-Materialisierungen*. Berlin: Lit Verlag.

Chen, Mel Y. 2012. *Animacies. Biopolitics, Racial Mattering, and Queer Affect*. Durham, NC: Duke University Press.

De Latis, Teresa. 1987. *Technologies of Gender: Essays on Theory, Film, and Fiction*. Bloomington: Indiana University Press.

Foucault, Michel. 1996. "What Is Critique?" In *What Is Enlightenment?: Eighteenth-Century Answers and Twentieth-Century Questions*, edited by James Schmidt and Kevin Paul Geiman, 382–98. Berkley: University of California Press.

Freud, Sigmund. (1931) 1963. "Female Sexuality." In *Sexuality and the Psychology of Love*, edited by Philip Reiff, 194–211. New York: Macmillan.
Griffin, Susan. 1978. *Woman and Nature. The Roaring Inside Her.* New York: Harper & Row.
Grusin, Richard. 2015. *The Nonhuman Turn.* Minneapolis: University of Minnesota Press.
Guattari, Félix. 2000. *The Three Ecologies.* Translated by Ian Pindar and Paul Sutton. London: Continuum.
Habermann, Friederike. 2017. "Nicht nur Identität, sondern auch materielle Verhältnisse queeren! Oder; Warum Marx und Butler dasselbe wollen." In *Material Turn: Feministische* Perspektiven auf Materialität und Materialismus, edited by Christine Löw, Katharina Volk, Imke Leicht and Nadja Meisterhands, 171–186. Opladen: Barbara Budrich.
Hall, Sarah. 2016. *The Wolf Border.* London: Faber & Faber.
Haraway, Donna J. 2008. *When Species Meet.* Posthumanities, vol. 3. Minneapolis: University of Minnesota Press.
Haraway, Donna J. 2016. *Staying with the Trouble. Making Kin in the Chthulucene.* London: Duke University Press.
Haushofer, Marlen. (1963) 1990. *The Wall.* San Francisco, CA: Cleis Press.
Herbrechter, Stefan. 2014. "'Nicht, dass ich fürchtete ein Tier zu werden...': Ökographie in Marlen Haushofers *Die Wand.*" In *Animal Traces/Tierspuren/ Traces Animales*, edited by Manuela Rossini. Vienna: Böhlau. (*Figurationen. Gender-Literatur-Kultur* 01/14: 41–55).
Horn, Eva, and Hannes Bergthaler. 2019. *The Anthropocene: Key Issues for the Humanities.* London: Routledge.
Knörer, Ekkehard. 2016. "Außer sich. Wo IT war soll Tier werden: 'Wild' von Nicolette Krebitz." *Cargo. Film, Medien, Kultur* 29: 4–7. https://www.vdfk.de/ausser-sich-691.
Krebitz, Nicolette. 2016. *WILD - Nicolette Krebitz zu Gast in Stuttgart.* Interview by Wolfram Hannemann. April 14. Video, 16:07. https://www.youtube.com/watch?v=H6pRQJJ0e9A.
Latour, Bruno. 2005. *Reassembling the Social: An Introduction to Actor-Network-Theory.* Oxford: Oxford University Press.
Lawrence, David H. (1923) 2006. "The Fox." In: *The Fox/The Captain's Doll/The Ladybird*, edited by Dieter Meh. London: Penguin.
Lettow, Susanne. 2018. "Die Stofflichkeit von Herrschaft. Materialismus als feministische Gesellschaftstheorie." In *Kapitalismuskritische Gesellschaftsanalyse: Queer-feministische Positionen,* edited by Katharina Pühl and Birgit Sauer, 241–257. Münster: Westfälisches Dampfboot.
Meißner, Hanna. 2014. "Von der Romantik imaginärer Verluste: Bringing the Material Back in?" *Femina Politica* 23 (2): 106–115. https://doi.org/10.3224/feminapolitica.v23i2.17618.
Minh-ha, Trinh. T. 2010. *Elsewhere, Within Here: Immigration, Refugeeism and the Boundary Event.* New York: Routledge.
Oshima, Nagisa. 1986. *Max, Mon Amour.* France/USA.
Preciado, Paul B. 2008. *Testo Junkie: Sex, Drugs and Biopolitics in the Pharmacopornographic Era.* Translated by Bruce Benderson. New York: Feminist Press.
Prokic, Tanja. 2017. "Becoming Wolf or, the Art of Affection in Nicolette Krebitz' *Wild.*" *Transfers* 7 (3): 131–134. https://doi.org/10.3167/TRANS.2017.070312.

Schadler, Cornelia. 2017. "Widerständige Apparate: Was ein anti-dualistischer und anti-dialektischer Materialismus zur Analyse von Differenz und Ungleichheiten beitragen kann." In *Material turn: Feministische Perspektiven auf Materialität und Materialismus*, edited by Christine Löw, Katharina Volk, Imke Leicht, and Nadja Meisterhands, 171–185. Opladen: Barbara Budrich.

Seier, Andrea. 2014. "Die Macht der Materie. What Else Is New?" *Zeitschrift für Medienwissenschaft* 6 (11): 186–191. http://dx.doi.org/10.25969/mediarep/1295.

Showalter, Elaine. 1981. "Feminist Criticism in the Wilderness." *Critical Inquiry* 8 (2): 179–205.

Tsing, Anna Lowenhaupt. 2015. *The Mushroom at the End of the World: On the Possibility of Life in Capitalist Ruins*. Princeton, NJ: Princeton University Press.

Varda, Agnes. 1985. *Vagabond* (original: Sans toit ni Loi) Oshima, Nagisa. 1986. Max, Mon Amour. France/USA.

9 Creating Emotion with Space in Nanouk Leopold's *Brownian Movement*

Angelica Fenner

Engendering New Languages of Trauma

In trauma studies it is maintained that extreme sensory encounters can radically upend familiar experience and thereby also defy cognitive processes of assimilation and ensuing narrativization.[1] Traumatic experiences become dissociated from conscious awareness and voluntary control, instead transferring to the subconscious. The challenge for storytelling, whether in moving images or in literary forms such as prose or poetry, then becomes how the unrepresentable is to be – if not represented – then, alternately, conjured or evoked. In what follows, I explore the possibilities via a Belgian arthouse film, Nanouk Leopold's *Brownian Movement* (2010) that participates in a broader movement in visual culture engendering new languages of trauma. By eliciting a haptic mode of viewing, one in which trauma is apprehended via sensory and affective engagement, Leopold's film unsettles traditional channels of identification with film characters and, by extension, any speculation about causal links between their present actions and possible unnarrativized or unnarrativizable pasts. As art historian Jill Bennett maintains,

> affect is not pre-coded by a representational system that enables us to read an image as 'about trauma,' then to experience it as secondary [i.e., second hand] trauma. The autonomic responses induced by affective imagery are, in their instantaneity, outside representation, [and] by the same token, they are not inside trauma.
>
> (Bennett 2003, 452)

Building on this conceptually rich and simultaneously slippery distinction between inside and outside, I argue that the spatialization of affect through film form can effectively conjure tensions between trauma's interiority, conceived as psychically and corporeally embedded, and the exteriority of memory, understood as trauma's secondary narrative reworking through artwork, interpersonal relations or engagement with immediate environs.

In *Brownian Movement*, trauma is limned via aesthetically striking natural and built spaces, while navigating between indoors and outdoors,

DOI: 10.4324/9781003023319-13

bodily cavities and corporeal surfaces, and (interior) psychical states made externally legible both through bodily affects and codified emotive expression. The spatio-textual poetics at stake are aptly captured in the term "architexture," originally coined by French literary scholar Gérard Genette ([1979] 1992). When grappling with the question of how to do literary theory after representationalism, and in particular, how the structuralist aspects of genre studies were to be understood or retained in representationalism's aftermath. Just as architectural structures emerge from a dialectic between structural form and open space (the latter associated with negativity or the void), Leopold's enigmatic film design emerges from interplay between its intertextual scaffolding, on the one hand, and representational voids and discursive fissures, on the other, while also foregrounding gendered dimensions in the enactment of its postfeminist, even posthumanist parable.

The plot revolves around a central female protagonist who appears to fulfil cultural expectations of doting motherhood, harmonious marital life and professional success. Charlotte von Ribeck, played by German actress Sandra Hüller, is a medical doctor from Berlin residing in Brussels with her husband Max (Dragan Bakema), an architect, and their well-mannered child, Benjamin (Ryan Bodie). Although the family occupies a spacious modern home surrounded by lush greenery, the story reveals early on that Charlotte has also rented a private flat downtown, where the spectator is soon privy to glimpses of sexual encounters with select male patients. Upon accidentally re-encountering one of these men on the construction site of a building her husband is overseeing, she suffers a mental breakdown, seeming not to recognize the man and instead succumbing to an inexplicable reflex to punch him in the face. We can presume this quickly brought her trysts into the public eye and to an ensuing end, for in part two of this tripartite film she is stripped of her medical licence by a review board and enters into marital counselling and psycho-pharmaceutical therapy. In part three, the plot fast forwards to Charlotte living in Ahmedabad, where her husband is overseeing restoration of an architectural heritage site while she cares for their growing family. Although Charlotte appears to have "settled down", her somnambulistic comportment implies repression of her past has come at a price. The film offers no clear resolution of the couple's arid relationship, and instead, suspends judgement through a sustained focus on aesthetic interiors sonically overlaid by a quixotic music score that holds viewers and protagonists alike in a state of languorous anomie sustained into the story's open ending.

My phenomenological approach reads the film's engagement with sensuality as irreducible to erotic or sexual frisson, and instead, as alluding to unreworked trauma. As elaborated above, trauma is only impoverished and distorted by representational frameworks, which reinscribe what physicist Karen Barad summates as "traditional empiricist assumptions concerning the transparent or immediate given-ness of the world" (2007, 152). In contrast to "the representationalist belief in the power of words to mirror pre-existing phenomena," it is the very nature of trauma to defy direct articulation or signification, as the originating events – singular or ongoing – explode the individual's ability to assimilate them in symbolic form and thereby contain or discharge altogether their unruly effects and affects (133). Instead, the psyche may defend itself through mechanisms of displacement and repression, such that trauma emerges, at best, as symptom, palpable or legible to a discerning observer *cum* spectator, therapist, or ally.

Although my reading of Leopold's film is not fully disentangled from psychoanalysis and the logic of the symptom, it is anchored in a posthumanist understanding of performativity indebted to Barad's theorization of what she terms "relational ontologies". Barad's thought retraces to Nobel Prize laureate Niels Bohr's quantum field model of philosophy-physics, which rejected the inherited Cartesian epistemology of mind/body, subject/object, and knower and known things mediated by language. Building on Bohr's rejection of "things" as ontologically determinate entities with boundaries and properties, Barad seizes as the unit of measure "phenomena", understood as

> [...] *the ontological inseparability/entanglement of intra-acting 'agencies.'* That is, phenomena are ontologically primitive relations – relations without preexisting relata. The notion of intra-action (in contrast to the usual 'interaction,' which presumes the prior existence of independent entities or relata) represents a profound conceptual shift. It is through specific agential intra-actions that the boundaries and properties of the components of phenomena become determinate and that particular concepts (that is, particular material articulations of the world) become meaningful.
>
> (Barad 2007, 139)

Leopold's film title already signals that quantum physics holds clues to the script's exploration of relationality. "Brownian movement" was first observed and identified by Scottish botanist Robert Brown in 1827 while using a microscope to examine pollen immersed in water. The term references the irregular and discontinuous motion of microscopic particles suspended in liquid or gas, which is caused by the particles' collision with faster moving molecules of the medium. Leopold's scalar shift from quantum particles to human "intra-actions" effectively captures the immersive spatialization of social relations (i.e. of relata), whose affective ecologies gain gendered contours under her masterful direction.

Haptic Film Design

Of her *oeuvre* as a whole, Leopold has remarked that she wants to be remembered "for creating a language in film that's my own. I always try to create emotion with space" (Exberliner 2011). Accordingly, in *Brownian Movement* architectural and environmental forms are deployed to strategic effect: pristine surfaces blend natural and synthetic materials seamlessly to connote controlled efficiency and sterile aesthetic harmony, whether through the glistening rows of glass vials and beakers in Charlotte's laboratory, the muted voices and crisp white walls of the hospital, or the warm tones and wood interiors of her domestic space. The film reveals the fissures in Charlotte's perfect life only gradually, but we recognize with hindsight that they were already evident from its opening moments, which encapsulate the film's overall audiovisual design. The credits open on diffuse unfocussed hues of pale, pinkish brown, while acousmatic laboured breathing reinforces a sense of psychical interiority. Ambiguously codable as either repressed weeping or, perhaps, orgasmic gasping, the sounds can also be understood as the exhalations of someone beginning to release, indeed, birth something new into being – not a child, but a new "sensory motor schema," to borrow Deleuzian terminology. The ensuing cut to an identifiable image, a sparsely furnished room, snaps the viewer out of the realm of sonic interiority to one of sonic exteriority via an extra-diegetic music score whose searching, even darting, melodic fragment will resurface at central points of the storyline without harmonically resolving.

These two opening images inaugurate a form of haptic visuality, which, according to Laura Marks' by now canonical definition, has

> the effect of overwhelming vision and spilling into other sense perceptions. This is, in part, because they [the images] do not provide enough visual information on their own to allow the viewer to apprehend the object, thus making the viewer more dependent on sound and other sense perceptions.
>
> (Marks 2002, 133)

True to this tactile and kinaesthetic approach, we encounter this second, focussed image as clean lines, smooth surfaces, and neutral colours comprising a domestic interior (Figure 9.1).

The space is frontally presented to the camera and subdivided via two archways framing entry into three successive chambers whose depth of field draws our gaze into an empty space of pure potentiality. Following a cut to black screen and white text mutely declaring "Part I," the woman we will come to know as the main protagonist enters frame right and slowly walks through the apartment, her gaze running along the walls, up the ceiling, towards the window, exemplifying how, in the words of Laura Marks, "the eyes themselves function like organs of touch" (2002, 162). Paraphrasing Aloïs Riegl, Marks maintains, this mode of looking, "tends to move over the surface of its object rather than to plunge into illusionistic depth, not to distinguish form so much as to discern texture," and "more inclined to graze than to gaze" (ibid.). When a woman's voice beyond the frame queries, "C'est

Figure 9.1 Charlotte von Ribeck's private flat (*The Brownian Movement*, dir. Nanouk Leopold, 2010).

un peu trop grand?" ("Is it too big?") Charlotte appears visibly perturbed by this disruption of her own focussed apprehension of a space she seems intent on *feeling* rather than assessing according to any practical criteria. Setting herself down on the bed behind her as if still in a trance, she responds to the proprietor's ensuing quip "the mattress is new," with another dumbfounded, even irritated, glance towards the voice off-frame.

We gain further clues to this haptic film design after Charlotte leases the space and occupies it for the first time. The sound of running water just off-screen saturates another static gaze, this time upon a bathroom lined with pinkish-tan, veritably nude-coloured tiles. The initial lack of visible human activity enacts a synaesthesia in which the tiled walls themselves become viscous membranes seemingly the conduit of fluidity, an organic extension of Charlotte body as she moves into the empty frame to dry her pale pinkened backside with a white towel. Hereby circumscribed is a space of both architectural and figurative invagination, such that what was perceived to be exterior proves interior, what was peripheral, becomes central, in a manner analogous to anatomical orifices such as the vagina, rectum, mouth, or ear, whose folds could be said to be both exterior and interior. The ensuing shot within the apartment reverses our earlier gaze through the three sequential chambers, now offering a lateral view of Charlotte in a silk robe standing motionless in the middle room, the bedroom. Approaching the capacious bed with its flesh-toned bedcover, she turns to seat herself, loosens her robe and lays down with legs dropping over the edge of the bed. Her bared lower torso is precisely aligned with a ray of sunshine beaming from frame left directly onto the bed, seemingly inviting the ray into her womb, of which the surrounding chambers now also become an architectural extension, their light-infused pinkish-tan walls muffling the exterior hum of urban activity and creating a soothing sound bath.

This downtown Brussels flat may appear an indulgent "supplement" to Charlotte's orderly life otherwise encompassing the gleaming modern, hospital and a charming, well-appointed home, but she arguably relates to it as a veritable encomium offering vital nourishment. The closing shot to the scene reinforces the analogy between womb and sheltering architectural interiors by positioning the camera at the backend of the building's ground-level public passageway connecting an inner courtyard to the front entry. A figure recognizable as Charlotte briskly descends from the flanking stairwell frame left into this dark interior passageway so akin to a birth canal, striding away from the camera towards the distant portal of light and street noise audible just beyond.

Invagination and the Intertext

Far from attributing the film's *mise-en-scène* an essentialist femininity, I would also point forth the centrality of invagination to its textual operations as a whole by drawing inspiration from philosopher Jacques Derrida's spatialization of textuality as folds or "pockets" of discourse, including his evocative analogy between the literal folds of book pages and those of the vulva. He also applies the concept of the fold, or invagination, to genre theory, conceiving of generic conventions as pre-existing structuring devices introduced from without, which thereby fold in on and become indistinguishable from (indeed, inherent to) the text as such (Derrida 1980). Invagination applies both to individual texts and to genres, such that the folds of any given text create "an internal pocket larger than the whole". This results from affinities of form and content that are shared across multiple texts to comprise a genre, but also from the fact that any given text is necessarily also shaped by influential texts that preceded it and that have infolded upon it, while, in turn, also influencing future texts and thereby undergoing redefinition of its own previous significance (Derrida 1981, 206).

Brownian Movement enacts this through several literary and filmic intertexts, each dramaturgically vital to the film's discursive operations. Perhaps most explicitly referenced is the Grimm fairy tale, "The Girl Without Hands," which Charlotte reads aloud to her son Benjamin at bedtime in a scene that immediately follows her first solo visit to the rental flat. We are only privy to her reading of the tale's happy end, recognizable through passing mention of the queen's iconic "silver hands".[2] Extensive critical scholarship on this particular fairy tale frames it as alluding to gendered processes of maturation or, even of incest and abuse. As such, it offers clues to the film's overall discourse on embodied female subjectivity and trauma while also foreshadowing the unravelling of Charlotte's marriage and its ambivalent resurrection. A tale as "Grimm" as they come, the original plot exists in different iterations that Wilhelm Grimm reworked, synthesizing disparate folk antecedents.

Regardless of plot variations, the widely known *ur*-tale adheres to a tripartite structure. Part one encompasses the miller's (failed) attempt to barter his daughter's life to the devil, his ensuing amputation of her hands, and her resulting decision to leave home, declaring, "Here I cannot stay". Part two traces her survival in the forest, discovery by a king, their marriage and birth of child, the Devil's second betrayal, banishment from the kingdom, and regeneration of her hands in the enchanted dwelling. Part three brings closure through the king's discovery of the forest dwelling, the couple's reunion and return to the kingdom to "live happily ever after". Leopold's film formalizes this tripartite division with intertitles announcing "Part One" (and so forth), but her postmodern rendering draws only figurative inspiration from literal events and takes liberties with their sequencing. Indeed, the debt to Grimm only becomes explicit through reference to the glass hands in the bedtime story, specifically excerpting the recuperation of the marital couple in the tale's denouement and thereby framing Leopold's own adaptation as critical postfeminist parable on the neoliberal terms under which women's autonomy is often voluntarily, when not by necessity, bartered for the illusion (the fairy tale) of "having it all" (e.g. career, family, relationship). Skirting the Grimm Brothers' earlier plot elements of parental abuse, bodily harm and flight from home, Leopold's adaptation instead brings forth affective qualities associated with the handless maiden, not least through the fairy tale's spatialization of trauma, which rendered the enchanted forest a space for physical and psychical regeneration in response to impossible demands made upon her body and psyche. Charlotte's rental flat similarly serves as provisional refuge from the radical violence of a socially enforced split between rationalized career life and a private sphere pressured to compensate for the resulting contradictions. The subjugation of the miller's daughter finds its contemporary corollary in modern women whose needs and desires remain misunderstood despite societal ideologies of sexual freedom and corporeal autonomy indeed, trauma's repression is central to the masquerade of feminine fulfilment widely espoused by postfeminist ideology.

Dissociation as Trauma's Spatialization

The plot elements constitutive of Part I of Leopold's film – the sexual encounters in the rented flat, the mental breakdown, and ensuing marital crisis – are suggestive of someone who may have *previously* survived a form of extreme trauma such as that elaborated in the Grimm tale. Jungian analyst Donald Kalsched maintains the original Grimm tale "illustrates how early emotional trauma can "freeze the bud" of a person's developing life and prevent it from flowering through an inner process of dismemberment and negation known as *dissociation*" (2013, 282). Individuals with dissociative behaviours experience a disconnect between the body, their thoughts and feelings, their surroundings, their actions and their identity. The disorder originates as an involuntary coping mechanism in response to the memory of overwhelming psychical and/or bodily pain experienced at a developmental stage when the individual lacked the defences to assimilate the experience. While protecting the coherence of the psyche, disassociation can also cause disruptions to functioning in everyday life.

Charlotte's comportment bespeaks similar dissociative states, exemplified in her aforementioned haptic mode of viewing the rental flat and ensuing trance-like interactions with strangers therein. She moreover seeks out men whose bodily characteristics dramatically contrast with those of her quintessentially handsome, svelte spouse. The patients who draw her attention during her medical rounds through the examination rooms of the hospital, and who later resurface in her flat, are alternately balding, extremely hairy, corpulent, or elderly. While the grounds for attraction remain ambiguous, one could speculate that, as a composite, their traits replicate those of someone who molested her as a child or young girl. Some of the scholarship on "The Girl without Hands" has framed it as a cautionary tale in which the Devil's lust for the girl serves as narrative displacement of the father's own incestuous desire for and ensuing violation of his daughter – a bodily violation that the "children's tale" displaces onto her hands, whose dismemberment, in turn, also metonymizes her dissociative response to the shock. Read thusly, the ensuing fairy tale becomes a survival guide for young girls in recovery, offering a narrative of social uplift and regrowth of hands that facilitates a "re-membering", understood not as memory work per se but as the psychological restoration of the integrity of psyche and its boundaries, and by extension, successful social assimilation into heteronormative frameworks of marriage and family. In a clinical context,

> dissociation does not occur as a result of external events alone. While it may appear to begin as an automatic response to unbearably painful events in a child's life, it is *kept alive* by inner factors or defensive powers that haunt or *hollow* the unconscious mental life of the trauma survivor... Sometimes patients will report feelings of vacuous emptiness inside, as if their bodies have been hollowed out like a shell
>
> (Kalsched 2013, 285–286; emphasis added)

The therapeutic process is one in which "[m]ind and body are slowly knitted together" to achieve what child analyst David Woods Winnicott captures in the term "indwelling of the psyche in the soma," a sense of bodily interior and exterior incorporated within a psychical reality ([1964] 1989, 113).

Following such a reading, Charlotte's retreat to a secluded urban refuge, with chirping birds audible through an open window, mimics the maiden's voluntary self-exile in the forest following loss of her hands. For both figures, this move is an act of self-preservation. Charlotte's haptic negotiations of the space recall the description of the handless maiden navigating the countryside without the use of her hands and instead using her mouth to pluck the fruit of a pear tree as do many non-human animals. Charlotte's sexual encounters similarly deemphasize use of her hands; rather, it is her skin as sensate organ that interfaces with the men she encounters, including when she passively submits to penetration of her various orifices. Derrida's generative understanding of invagination spatio-temporalizes the posited psychodynamics, in which Charlotte experiences this chamber as an infolding of her own bodily cavities past and present, into which she voluntarily receives strangers. Even if she thereby potentially restates prior experience(s) of perhaps involuntary violation as instead non-shaming voluntary supplication, these encounters are not reducible to the trope of "sex therapy". Instead, I surmise Leopold aims for us to apprehend them as "intra-actions" instantiating boundaries anew amid "matter in the making," rather than reinforcing existing boundaries between already formed and fixed bodies. What renders the scenes in Charlotte's chamber unintelligible within a representational paradigm is that they are affectively coded neither as pleasurable nor as painful. Depicted instead is a form of wayward agency and constitutive openness in enigmatic scenes of bodily intimacy that enact a folding, in which neither "outside" nor "inside" (both corporeal and environmental), understood here as the minimal determinants of "form", can be read as stable or self-identical.

Touching Tableau: *Belle de Jour's* Legacy

Charlotte's forays gain further significance from Luis Buñuel's new wave classic *Belle de Jour* (1967), identified by Leopold as also "a great inspiration" (Exberliner 2011). Buñuel's adaptation of Joseph Kessel's (1928) erotic novel depicts upper-class French housewife, Séverine Serizy (Catherine Deneuve) "working" several afternoons a week at a discreet high-end brothel downtown. Buñuel never renders explicit the motivation for her extra-marital forays from her spouse Dr. Pierre Serizy (Jean Sorel), a Count who is also physician at a hospital. Yet affective facial cues in Deneuve's performance of a bodily comportment described by more than one scholar as "frozen" imply Séverine enjoys being forced into submission most especially by clients of proletarian background.[3] That masochistic preference, in turn, gains potential significance when she recalls in passing to a brothel co-worker being "touched" by mechanic when she was eight years old, a comment followed by a fleeting image of a young girl in a night gown embraced by a man in blue coveralls. Buñuel effectively dabbles in psychoanalytical allusions without fully committing to them, inserting throughout the film similarly abrupt tableaux shots temporally and causally ambiguous in their status as objective or subjective flashbacks of her past, or alternately, as masochistic fantasy.

Nanouk Leopold maintains her film "has a similar subject but in another time with other values" (Exberliner 2011). The idle and also childless housewife Séverine gets recast for the neoliberal era as Charlotte, exemplar of similar privilege who appears, on first glance, to fulfil normative ideals of family life and of a successful career – the latter, by usurping the medical discourse and professional authority Buñuel previously assigned Count Serizy. Charlotte's work as a physician is first introduced in a strikingly formal scene: a languid lateral tracking shot of her performing her rounds in a clinic, walking slowly past rows of partially curtained-off spaces. The mood is hushed, the ubiquitous white curtains dampening the soft murmurs and rustle of clothing. The sharp, steady sound of Charlotte's heeled shoes treading the floor lends her mobilized gaze aural contours and signals her authority, punctuated when she pauses upon glimpsing a seated man rolling up his sleeve for a nurse. When the man turns to meet Charlotte's gaze, she steps into the cordoned space and offers to "take over from here". Which she does, indeed, following a match cut from Charlotte frame right looking upon the seated patient before her to a shot of Charlotte again frame right standing immobile in her bathrobe, now in her private suite, gazing leftward as a door is heard opening and closing off frame, and, to the spectator's surprise, the same man enters in his street clothes. Wordlessly, he steps forward to stand close to Charlotte and reaches with one hand for her jaw, clutching her mouth open as he inhales her breathe in a detached, but inquisitive manner that mimics but also reverses the direction of the physician's previously glimpsed clinical bodily assessments.

Charlotte's privately orchestrated affairs appear like a more prurient mode of medical reconnaissance, yet it is she who submits to the probing of others in deterritorializing encounters that accord with Barad's conceptualization of agential realism. Rather than two pre-existing individuals entering into codified patterns of relation, these screen performances explore how differences – here, along multiple axes of power, gender, class, bodily shape, size, texture – "are made and remade, stabilized and destabilized, as well as their materializing effects and constitutive exclusions" (Barad and Kleinman 2012, 78). Whereas Buñuel's brothel scenes and abrupt jump cuts coded as sado-masochistic fantasy or memory conjure tableaux of sexualized submission along classical psychoanalytical pathways, Leopold queers the familiar sense of causality (where one or more causal agents precede and produce an effect), and more generally unsettles "the metaphysics of individualism" (77) to suggest that things in the material world exist not in themselves but within relations, understood as "differential patterns of mattering [...] dispersed across differently entangled spaces and times" (78).

To think matter, as Barad does, as "substance in its iterative intra-active becoming – not a thing but a doing, a congealing of agency" finds its philosophical corollary in Leopold's environmental film design (Barad and Kleinman 2012, 80). For example: Charlotte's later private encounter with a corpulent hairy man is introduced via cropped close-up of a woman's fingers tracing folds and creases of flesh along a round, hairy torso, and then reaching with both hands to cup and lift the rotund belly, the implied weight of its fecund form perhaps raising the memory of her own once-pregnant belly. No words are exchanged in a scene whose sole Foley-enhanced sounds include skin brushing skin and air entering and leaving the lungs (which themselves arguably comprise a filigreed invagination of countless epidermal surfaces intra-acting with molecules of oxygen, hydrogen and carbon). The camera pulls back to reveal that their nude bodies rest on a soft white shag rug whose texture, analogous to the hairy surface of the man's skin, invites a visceral spectatorial response analogous to that elicited by ASMR (Autonomic Sensory Motor Response) videos. The next formal shot presents Charlotte laying on her belly on the same rug, facing the camera with left arm propped and the right extended sideways to luxuriate in the rug's tickling fibres; in turn, the man's massive torso silently re-enters the frame from behind Charlotte as he crawls on all fours atop her backside, his face remaining obscured by the upper film framed even as he slowly lowers his massive weight with a long and heavy sigh.

This intra-active reconfiguring of boundaries through shifting points of contact between textured surfaces, in which epidermal and environmental become virtually synonymous, poses only one of several scenes that deconstruct the very supposition of figure and ground. Whether with strangers or within the familial triad and marital dyad, these encounters establish Charlotte's relationship to different modes of intimacy as non-hierarchical, perceiving love of family and spouse as on a continuum with, even complementary to the pleasures experienced in other relational ontologies whose boundaries she does not seem take for granted (Barad 2007, 91). One can only speculate that past trauma has reconfigured her sense of where "agential cuts," or local resolutions in ontological indeterminacy, are to be drawn. These current "therapeutic" encounters are now motivated by a desire to investigate anew the material-discursive practices that enact these boundaries, and with it, produce subjects and objects and other attendant differences.

The Biopolitics of Hospitality and Care

Charlotte eventually becomes culpable for crossing the ethical threshold set into place by a medical apparatus strictly regulating the divide between subject/object, observer/observed, knower/known, such that the physician is accorded the status of subject and authority, while the patient consents to an objectification imputed to serve collectivized interests of biopolitical regulation. Those disparate regimes of bodily (self)care, that is of respectively medical treatment and of consensual private intra-relation, collapse when she unexpectedly encounters on her husband's work site a construction worker, also her patient, with whom she previously met in her private flat. Perhaps unable to cognitively integrate the decontextualized encounter, her visceral reflex is to punch him in the nose, followed by the equally swift professional imperative to reach out in concern, only to assault him again in schizoid confusion before collapsing and losing consciousness. Part

1 closes with her lying in a hospital bed, her husband quietly watching her nearby, while Part 2 commences with counselling and psychopharmaceutical treatment.

The palatial office of the woman psychiatrist whom Charlotte visits is architecturally striking: its expansive exterior wall curves outward with a series of oversized floor-to ceiling windows, permitting natural light to saturate the opposing interior white walls with a sterile luminosity. Devoted to a rationalist understanding of the workings of the human psyche, this clinical-therapeutic space overdetermines the very type of subject it purports to treat – rendering Charlotte small, forlorn, and disoriented in striking contrast to the soft light infusing the private chambers of Charlotte's earlier corporeal encounters. The "stage" blocking alone conveys a lack of mutual congress: the psychiatrist stands beside her desk absently gazing down at some paperwork with her back to Charlotte, who slowly wanders the expansive room seemingly at a loss for words (Figure 9.2).

The talk therapy fails to bring clarity and Charlotte's lines of dialogue signal a cognitive resistance to explaining her past behaviours. When the psychiatrist asks, "How is it going with the new medication?" the camera cuts to Charlotte transfixed before a large woven wall tapestry depicting a primeval forest – one that recalls the queen's seven years of exile in the Grimm fairy tale. Her back to the camera and the doctor, Charlotte replies, "I really don't know what I'm supposed to feel" while scanning the tapestry as if for clues. Birdsong from beyond an open window now gains sonic focus, as if emanating from the fabric's own acoustic depths and recalling the same ambient sounds faintly audible in the rental flat during the opening sequence. Charlotte's broader psychical journey in Part II herein becomes one of wandering anew through the Grimmian enchanted forest. Recall the Queen's second journey into the forest is not in search of immediate safety as was her first flight following amputation of her hands. Leopold's film locates any

Figure 9.2 A lack of mutual congress at the psychiatrist's office (*The Brownian Movement*, dir. Nanouk Leopold, 2010).

earlier loss or violation prior to the diegetic timeline while circumscribing its precise origins, which remain an unspoken and unspeakable trauma. If the first journey in both the fairy tale and Leopold's inspired adaptation entails self-imposed flight, the second redoubles the earlier literal amputation through forcible excision from the socius followed by eventual rehabilitation, entailing, in Charlotte's case, concerted efforts at gender conformity.

Subtly stencilled after the Grimmian tale, Charlotte's hands constitute ciphers both of agency and of the terms of environmental and social mediation in an individual with dissociative tendencies. This is exemplified immediately following her second session with the psychiatrist. Charlotte stands alone outdoors, dazed and disoriented. Behind her, the backside of an industrial loading dock, and beside her someone's failed attempt at urban landscaping: several oversized poured-concrete containers housing neglected plants and grasses, clearly parched and rewilding. She seats herself on the edge of one these structures, becoming just another abjected life form deemed out of place within the existing constraints. Very little happens in this durational shot lasting fully three-and-a-half minutes with only one cut. Instead, the camera holds on subtle affective shifts legible in Charlotte's comportment and her visage as she gradually melds with her surroundings, again enacting a redistribution of sensible preconceptions of figure and ground, which, in Jane Bennett's words, would otherwise "divide the world into dull matter (things) and vibrant life (us, beings)" (Bennett 2010, vii). This division, which "ignores the vitality of matter itself" gradually dissolves as Charlotte falls into another sensorial reverie attuned to a hybrid nonhuman symphony composed of chirping birds, the distant hum of traffic, a car horn, a passing train rattling along its tracks, a police siren's brief howl, the vigorous chattering of crows. Clearly lost in thought, her facial expressions register shifts in her "aesthetic-affective openness to material vitality" (x); she eventually looks down to contemplate her hands, holding them out before her as if viewing them for the first time in a gesture that would mystify any spectator who doesn't recognize the link to the titular motif of "The Handless Maiden".

Charlotte's "strange combination of delight and disturbance" (Bennett 2010, xi) signals a cognitive dissonance between her personal ethics and that of the Central Medical Tribunal where she is next summoned for a hearing. The medical profession constitutes part and parcel of, in Foucault's terminology, a "dispositif", or apparatus, defined as "a thoroughly heterogeneous ensemble consisting of discourses, institutions, architectural forms, regulatory decisions, laws, administrative measures, scientific statements, philosophical, moral and philanthropic propositions≈– in short, the said as much as the unsaid" (Foucault [1977] 1980, 194). Its operations are affectively conjured in the architectural building housing the Tribunal, whose interior, not unlike the psychiatrist's office, is characterized by imperiously tall windows and doors at least twice the height of the humans permitted to pass through them, their dark wood matching the panelled walls to convey

Figure 9.3 Charlotte awaiting trial in the hall outside of the Medical Tribunal (*The Brownian Movement*, dir. Nanouk Leopold, 2010).

lasting strength. This institutional authority dwarfs Charlotte, seated on a hallway bench outside the courtroom awaiting her hearing.

Trapped in the cross-hairs of differing ethical imperatives, her own and those of the Tribunal, Charlotte's predicament enacts a peculiar metonymy with the huge portrait painting of the biblical Jesus looming on the towering wall above her (Figure 9.3). Jesus was known for extending compassion indiscriminately, thereby defying the caste system of his time and placing its social hierarchies so radically into question that he was put on trial. Analogously, Charlotte's manner of "hosting strangers" in her private suite takes to another level entirely the terms of compassionate care mandated by her workplace, the hospital, whose Latin cognate, *hospes* translates as "host" and late Latin *hospitale* as guest house. The ensuing court scene formally captures this conflict of identifications through a posterior view of Charlotte's head as she faces the tribunal visible within the depth of field, thereby positioning her between the reticles of respectively their judgemental and our empathetic gaze as she awaits the impending ruling.

Under interrogation is Charlotte's agency and culpability as a physician, or in Barad's language, where she and her interlocutors, in this instance, her (male) patients, draw agential cuts. Her world-making seems aligned more with the non-predeterminate objectivity extrapolated from the realm of quantum physics, such that relata (here, doctor patient relations) "do not pre-exist relations; rather, relata-within-phenomena emerge through specific intra-actions" (Barad 2007, 140). These entanglements obviously defy the linear causalities underpinning the Cartesian *dispositif* of the Disciplinary Tribunal and as such, Charlotte's behaviours also put into question the ethical distinction drawn between the social and the scientific. They moreover interrogate female agency in particular, evidenced in the pointed irony of a woman judge flanked by other (female) court functionaries, who conveys the jury's serious doubts about Charlotte's capacity to execute her duties in a satisfactory manner; in particular, the judge cites her failure to comprehend the responsibilities of her profession, a lack of vision, and an underdeveloped conscience.

In a reverse close-up, Charlotte remains impassive if not defiant, her lack of regret for her behaviours signalling that she simply could not do otherwise, even as silent tears roll down her face at the harsh judgement. Her banishment from powerful institutional privileges mimics Part II in the Grimm tale, that is the renewed exile of the Queen without hands after the Devil intercepts her letter to the king and plants the false pronouncement that she has birthed a monster. Whereas the fairy tale locates the imputed freakery in a woman's fecundity, Leopold's rendition locates it in non-reproductive sexual relations socially codified as perverse. Indeed, when previously asked by the psychotherapist to identify what distinguishes the sexual encounter with Max from those with strangers, Charlotte volunteers, "Max is a very handsome man, he's attractive and I want to touch Max always, that's almost automatic". When the therapist clarifies, "You mean you didn't really want to touch the other men?" she begins and then trails off, "No, yes, I did, but it's a completely different sort of touching, it gives me an entirely different [...]" She confirms she did pick out the men for certain physical traits (bitten nails, pock-marked face, hairy body, corpulence), but can't or won't explain why, adding, "I think I shouldn't be saying it. I shouldn't tell at all. It only makes it worse". Charlotte voluntarily identifies touch as a means of encounter, of which human hands are one mediating interface, serving also as instruments both of exploration and manipulation, equipped as they are with opposable thumbs. If her extra-marital bouts of touching others violate the matrimonial terms that limit touching exclusively to one designated body, they also transgress the codified terms of touching sanctioned by the medical profession. When the judge formally revokes her licence out of the Tribunal's concern that this has led to seduction, here understood as instrumental objectification of those bodies under her care, Charlotte's personal and professional "disarmament" redouble one another.

The Dialectics of (Dis)Enchantment: Wayward Women on Trial

On trial then, is not only Charlotte's professional integrity, but female desire and agency broadly defined, powerfully condensing multiple textual avatars. Charlotte's hair twisted into a bun and viewed from behind against the backdrop of the jury is strikingly reminiscent of the back of Kim Novak's head in Hitchcock's *Vertigo* (1958) viewed when her character Judy Barton – another woman living an illicit double life – sits in a museum viewing a painting of her nemesis Madeleine Elster. In that cautionary modernist narrative of human desire, desperation, and the entanglements of fantasy and reality, it is ultimately Judy's duplicitousness that is placed on trial for misleading the gullible Scottie. Sévérine in *Belle de Jour* also sports a pert chignon, but this genealogy retraces a step further to Buñuel's own professed debt to Lewis Carroll's classic nineteenth-century tale *Alice's Adventures in Wonderland* and *Through the Looking Glass,* whose eponymous heroine is characterized by a peculiar mixture of curiosity and social awkwardness replicated in Charlotte's quixotic character (Carroll [1865] 1982). Carroll's beguiling fantasy novels were already favourites among the surrealists of the 1920s for upending traditional causalities associated with that established and imposed social order known as "reality" (Duprat de Montero 2013). In Leopold's iteration, Charlotte's rental suite poses an analogous rabbit hole down which she slips into a trance-like state of enchantment, and like Alice before her, Charlotte eventually finds herself in a courtroom that enacts a form of justice incommensurable with her own moral sensibilities. Truly on trial in Carroll's cleverly disguised social satire for adult children is the legal system itself, just as Leopold's arepresentational approach interrogates the ethical frameworks that delineate agency and morality and does so not only diegetically but also by interrogating what can be conveyed within the cinematic *dispositif.*

Located shortly past the halfway mark in the film's 90-minute run time, the courtroom scene brings to a head the ethical tensions between what could be framed as "the disenchantment of modern life", associated with adherence to reason, delimited freedoms, efficiency and control, on the one hand and those of surrendering to enchantment, a somatic state in which, in Jane Bennett's words, one is "struck and shaken by the extraordinary that lives amid the familiar and the everyday" (2001, 4). Charlotte's medical profession demands fealty to an ethical code of conduct potentially at odds with the affective forces that otherwise inspire ethical choices – a conflict Bennet's reasoning resolves dialectically by advocating for "honing sensory receptivity to the marvelous specificity of things" according to the logic that "to some small but irreducible extent, one must be enamored with existence and occasionally even enchanted in the face of it in order to be capable of donating some of one's scarce moral resources to the service of others" (ibid.). Within the broader framework of trauma and memory work signalled in Leopold's intertexual homages, it can also be argued that trauma, the result of a shock that exceeds the individual's capacity to assimilate the sensory input, lives somatically very close to enchantment, "a sense of having had one's nerves or circulation or concentration powers tuned up or recharged," even as the origins obviously differ (5). Thus, Charlotte's seemingly "frozen", or somnambulistic comportment in the scenes of intimacy with strangers can be read to symptomatize earlier trauma even as it also signals a radical sensory openness.

In this dialectic of enchantment and disenchantment, it is instructive to recall that the Grimmian enchanted forest was also the locus for divine intervention: a guardian angel who assisted the Queen's hands to magically regenerate, enabling her to cast off the glass prostheses. Psychoanalytically speaking, areas of the self that had become hollowed out, "frozen" and fragile like blown glass lacking pliancy, are again infused with the pulse of life and a renewed capacity for productive exchange with the world. Leopold relocates the enchanted forest in Charlotte's private suite, a space facilitating an equally mysterious form of re-membering not limited to the hands but reactivation of the whole body as vital medium. Not merely the locus for this therapy, these chambers invaginate Charlotte's very psyche, facilitating its restructuring, analogous to the more literal processes of gastrulation in developmental biology from whence the term invagination originally hails.

The *Chora*, a Space of Becoming

Charlotte's claim upon this private suite and what transpires there correlates with what Derrida has described as "the logic of supplementarity," whose dialectic also underpins his concept of invagination ([1967] 1997, 144–145). That logic emerged from his study of the writings of Jean Jacques Rousseau, who overcame binary terms such as "nature" versus "society" through the concept of the *supplement*, understood as what gets added from the outside that provides something that was previously missing and that was thus, effectively not only necessary to that to which it now gets added but actually preceded it in origin (ibid.). Derrida demonstrates via Rousseau how the supplement enables the unity of otherwise opposing gestures, or allows contrary claims to cohabit, thereby overcoming what philosophy normally seeks to resolve, namely contradiction. Applying an analogous "double reading" to *Brownian Movement,* I have argued that Charlotte's spatialized "forays" in the rental suite follow this logic of supplementarity, effectively revealing the actual impossibility of the social contract, which Laurent Berlant delineates as the promise of upward mobility, job security, social and political equality, and durable intimacy to fill a lack that preceded those relationships, while also enabling her to endure that void and "live on" (Berlant 2011).

In the archetypal fairy tale, it was ultimately patriarchal society, broadly conceived, that inaugurated this trauma, positing those gendered female as the site of lack, not fully sovereign and reduced to objects of barter. Leopold's arepresentational realism, by contrast, correlates with an aesthetics of absence that is mapped spatially and does not code lack negatively. Rather, that aesthetics revaluates negativity as such, aligning not only with Derrida's understanding of invagination, but also of the *chora*, a central concept both in his philosophy and that of Julia Kristeva, understood to render space an expressive element that, however obliquely, conveys the inexpressible. For Kristeva, the *chora* as space of becoming

> acts across the organization (the structure or finitude) of the text in which the chora of the process is figured. This chora is the non-verbal semiotic articulation of the process; music or architecture might provide metaphors better suited to designate it than the grammatical categories

> of linguistics which it actually organizes. It is the logic of the 'concrete operations,' of the 'motility' [...] which runs through across the body in social space (transforming objects, relations to the family and to the social domain).
>
> (Kristeva 1998, 142–143)

In the third and final segment of Leopold's tripartite tale, a relationship between space, art, and female autonomy that has been building from the opening sequence is thus achieved as much through its construction of *textual space* as through its operations as *spatial text*, meaning that spatiality functions not solely or even primarily as "motif" but is the very scaffolding of the film discourse. In this regard, the arepresentational realism of her project as a whole could be said to instantiate the *chora* for the spectator.

Part Three opens on a radically different topography while also signalling a leap in time. The door on Charlotte's professional life has closed and she has accompanied Max to Ahmedabad as he oversees restoration of Le Corbusier's Mill Owner's Association Building. The contrastive topography and culture arguably intensifies via *photogenie* the phenomenological experience of the film's *mise-en-scene*, offering framed glimpses from within built spaces through windows, doors, or other built structures onto audiovisually vibrant and poetically nested spaces that are infused with striking qualities of light. Several times, Charlotte can be seen entering the cavernous interior of her husband's construction site to wander its empty corridors and wings, eventually sitting with her back against a pillar, asleep, or standing trance-like staring into space. Growing suspicious of her motivations, Max silently follows her and observes her at a distance from dark corners. Yet Max does not find what he continually expects, for what is at stake remains essentially irrepresentable. As an architect, he is trained in the construction of form, but in the process seems to overlook the void which every form opens and enables, while Charlotte for her part, retains an affinity most especially for that negative space. Just as the Grimmian King travelled seven years through the thickets of the enchanted forest before finding the Queen and returning with her to civilization to rebuild their lives and live happily ever, Max could be said to have drawn Charlotte away from her private flat in Brussels to launch a new life in another cultural setting. But Charlotte, while going through the motions of devoted mother and loyal spouse, has not surrendered that Derridean "negative space" of pure potential, one that enables an openness to re-enchantment even within modernity. Perhaps Leopold herein rewrites the implied close to the original fairy tale to imply that the Queen might have been perfectly content to remain in the enchanted forest rather than return to the civilizing influences that twice betrayed her.

The film's closing sequence leaves ambiguous whether Max remains destined to continue figuratively "searching the forest", haunted by his earlier memories of object-loss as he anticipates Charlotte's renewed disappearance. In the couple's penultimate interaction, he lies in Charlotte's arms one night and says "Sometimes I don't know who you are anymore". She replies, “I’m the same person,” to which he responds after a long pause, “I don’t know if that’s enough”. Her final words, “I’m here, now. Why can’t that be enough?” pose an existential query that remains unanswered. Their exchange aptly captures the viewing public’s often disparate responses to the film as a whole, whose negative aesthetics cannot concretely depict what is at stake and may simply not “be enough’ for some spectators, even as others, including the author of this essay, may be captivated, indeed, enchanted. The final setting opens on a subjective view through the window of an all-terrain vehicle moving across a light-infused arid desert landscape in which sky and land merge as one vast and empty open space. A variation on the eerie electronic soundtrack previously associated with the Brussels flat gradually returns, while the vehicle’s sideview mirror visible in the lower left corner of the film frame produces a time-image of simultaneously unfolding present and receding past. A reverse shot reveals Max in the backseat with one arm hanging from the grip above the window, and then Charlotte beside him, both gazing out their respective side windows into the desert. She turns to look in his direction, smiles, and leans sideways to rest her head upon his shoulder and close her eyes as the camera holds for another minute on the pair jostled by the chauffeured vehicle’s movements.

An open ending familiar from Berlin school films with which this Dutch production has also been compared, it locates both protagonists in a state of becoming in the wake of past trauma. The choice of setting reprises the film’s titular allusion to the random collision of molecules suspended in liquid, this time via a comparatively arid topological metaphor for protean immersion and haphazard movement: the desert terrain, conceived by Deleuze and Guattari as a “nomadic space”, where sweeping winds may cause defining markers to disappear, forcing the mobile subject to surrender any claim to a linear route along the dunes and instead pursue, at best, a trajectory or “line of flight” along its “smooth”, mutable surfaces ([1980] 1987, 380–384). Max and Charlotte alike appear to have adopted the status of nomads navigating a world whose convulsive ripples they can neither fully anticipate nor fathom as these striate the shifting sands of their relationship, or alternately, precipitate further collisions within flexible constellations of individuality, work, and community that characterize what Zygmunt Bauman has coined “liquid modernity” (Bauman 2000). As the closing credits return us to the unfocussed tan/pink hues of the film’s haptic opening shot, only short-sighted vision remains possible.

Acknowledgements

Early versions of this paper were presented at the invited conferences “Ecologies of Gender” at the Free University of Berlin in 2018 and “Feminism, Theory, Film” at the University of Oregon in 2019. For their vital feedback and editorial assistance, the author thanks respective conference organizers Sabine Nessel and Susanne Lettow (Berlin) and, posthumously, Sonia Boos (Oregon) as well as dedicated research assistants Sophie Holzberger and Erin Mick.

Notes

1 This has been most prominently debated in Cathy Caruth's edited volume *Trauma: Explorations in Memory* (Baltimore, MD: Johns Hopkins University Press, 1995) and her monograph *Unclaimed Experience: Trauma, Narrative, and History* (Baltimore, MD: Johns Hopkins University Press, 1996).

2 Grimm's *Das Mädchen ohne Hände* has been translated as *The Girl with Silver Hands,* but also as *The Girl with the Glass Hands.*

3 The adjective "frozen" similarly surfaces in the English translation of Kessel's novel when describing Sévérine's response upon hearing of another woman of her own social class voluntarily serving in a brothel: "Her face was frozen, caught in some invisible mold, unbreathing". Deleuze, in his commentary on the display of submissive sexual behaviour in Sacher-Masoch's *Venus in Furs,* similarly writes that they, "have of necessity a frozen quality, like statues or portraits" (Deleuze 1989 [1967], 69).

References

Barad, Karen. 2007. *Meeting the Universe Halfway: Quantum Physics and the Entanglement of Matter and Meaning.* Durham, NC: Duke University Press.

Barad, Karen and Adam Kleinman. 2012. "Intra-actions: An Interview with Karen Barad." *Mousse* 34: 76–81.

Bauman, Zymunt. 2000. *Liquid Modernity.* Malden, MA: Blackwell.

Bennett, Jane. 2001. *The Enchantment of Modern Life: Attachments, Crossings, and Ethics.* New York: Princeton University Press.

Bennett, Jane. 2010. *Vibrant Matter: A Political Ecology of Things.* Durham, NC: Duke University Press.

Bennett, Jill. 2003. "Insides, Outsides: Trauma, Affect, and Art." In *The Feminism and Visual Culture Reader,* edited by Amelia Jones. New York: Routledge.

Berlant, Lauren. 2011. *Cruel Optimism.* Durham, NC: Duke University Press.

Buñuel, Luis, dir. 1967. *Belle de Jour* Alliance Atlantis, 2003. DVD.

Carroll, Lewis. (1865) 1982. *Alice's Adventures in Wonderland and Through the Looking Glass.* Edited by R. Landelyn Green. Oxford: Oxford University Press.

Deleuze, Gilles. (1967) 1989. *Masochism: Coldness and Cruelty.* New York: Zone Books.

Deleuze Gilles and Félix Guattari. (1980) 1987. *A Thousand Plateaus: Capitalism and Schizophrenia.* Translated by Brian Massumi. Minneapolis: University of Minnesota Press.

Derrida, Jacques. (1967) 1997. *Of Grammatology.* Baltimore, MD: Johns Hopkins University Press.

Derrida, Jacques. 1980. "The Law of Genre." *Critical Inquiry* 7 (1) (Fall): 55–81.

Derrida, Jacques. 1981. "The Double Session." In *Dissemination.* Translated by Barbara Johnson, 173–285. Chicago, IL: University of Chicago Press.

Duprat de Montero, Arnaud. 2013. "The Intertextual Presence of Lewis Carroll's Alice in *Belle de Jour.*" In *A Companion to Luis Buñuel,* edited by Rob Stone and Julián Daniel Gutiérrez-Albilla, 479–493. Oxford: Blackwell Publishing.

Exberliner. 2011. "Nanouk Leopold: The Brownian Movement." (February 4). https://www.exberliner.com/whats-on/film/nanouk-leopold%3A-the-brownian-movement/.

Foucault, Michel. (1977) 1980. "The Confession of the Flesh." In *Power/Knowledge: Selected Interviews and Other Writings,* edited by Colin Gordon, 194–228. New York: Pantheon Books.

Genette, Gérard. (1979) 1992. *The Architext: An Introduction.* Translated by Jane Lewin. Berkeley: University of California Press.

Hitchock, Alfred, dir. *Vertigo.* 1958. Universal Studios, 2012. DVD.

Kalsched, Donald. 2013. *Trauma and the Soul: A Psycho-Spiritual Approach to Human Development and Its Interruption.* London: Routledge.

Kessel, Joseph. 1928. *Belle de Jour.* Paris: Gallimard.

Kristeva, Julia. 1998. "The Subject in Process." In *The Tel Quel Reader,* edited by Patrick French and Roland-Françoise Lack. New York: Routledge.

Leopold, Nanouk, dir. 2010. *Brownian Movement.* First Run Features, 2014. DVD.

Marks, Laura. 2002. *Touch: Sensuous Theory and Multisensory Media.* Minneapolis: University of Minnesota Press.

Winnicott, D.W. (1964) 1989. "Psycho-Somatic Illness in Its Positive and Negative Aspects." In *Psychoanalytical Explorations,* edited by C Winnicott, R. Shepherd, and M. Davids, 103–114. Cambridge, MA: Harvard University Press.

10 An Ecohumanist Perspective

Theorizing Ecofeminism through a Spatial Analysis of Amitav Ghosh's *The Hungry Tide*

Sangita Patil and N S Gundur

The aim of this chapter is to critically explore the representation of gender and environment in the Indian context through a close examination of Amitav Ghosh's *The Hungry Tide* (2008). By reflecting on the nature–culture dialectic, this literary enquiry makes us rethink the interrelationship between the human and the non-human in a way that blurs the boundaries between these categories and helps us to envision an ecohumanist perspective. This new perspective is based on epistemological, empirical, lived experiences of the gender–environment relationship, and critically responds to the ecofeminist project. Some stances of ecofeminism are still relevant such as ideas about the future planet, the fight against nuclear weapons, toxic waste, genetic engineering, and development. However, we depart from the core argument of ecofeminism the gender/women–nature nexus, and instead propose one based on the connection between humans and nature, under the heading of humanism. This proposal breaks with customary disciplinary boundaries and provides a broader, intersectional platform that focuses on the construction of human experiences at the intersection of both social and ecological systems (Thompson 2016). We understand intersectionality as an interplay of human and nature based on socio-spatial differences and materialities (Sultana 2020), Global North and Global South, class and caste, urban and rural, and so on.

Amitav Ghosh's *The Hungry Tide* emblematizes a diverse range of ecological narratives. Ghosh is one of the most erudite contemporary Indian writers, who has produced a corpus spanning a wide range of literary genres, including essays, fiction and non-fiction. Though his works deal with diverse themes and concerns, such as fundamentalism, cultural heritage and political/historical questions, his major focus is on the topics of ecological degradation and climate change which he tackles in his most acclaimed novels *The Hungry Tide*, *Sea of Poppies* (2015) and *River of Smoke* (2011). He dreams of a state of "biophilic mutuality" for all living beings, which could possibly lead to the establishment of a new ecological ethics and environmental justice. His widely acclaimed non-fiction work *The Great Derangement: Climate Change and the Unthinkable* (2016) is about

DOI: 10.4324/9781003023319-14

how human interaction with nature is leading to climate change and environmental catastrophe. It argues that we urgently need to reflect on our activities and amend them appropriately. Before this non-fiction work, Ghosh attempted to portray the ecological crisis and the importance of the flora and fauna of the Sundarbans in his iconic novel *The Hungry Tide*. The novel's dominant mode of narration is realism, which it deploys to reflect on the essence of India's environmental problems – that is to say, on contemporary reality – through a focus on the interactions between humans and non-humans. As Ghosh says, "I am greatly beholden for my earliest linkage of memory with the tide country" (2008, 429). As an anthropologist, he visited and studied the Sundarbans, and so the novel is tinged with a native's experiential ethos. Some of Ghosh's characters are based on real people – Isabel Beasley appears as Piya Roy, Tushar Kanjilal as Nirmal, Bina Kanjilal as Nilima, etc. These borrowings from real life not only draw on the historicity of the Sundarbans but interrogate a more universalist ecological position.

Applying Edward Said's method of contrapuntal reading to the novel helps us to understand a wide range of relations: power inequalities in general and the connectedness of humans with their natural environment in particular. One new way to conceptualize gender and environment is through the perspective of ecohumanism, which highlights the connectedness of humans with their natural environment. Ecofeminism is struggling hard against development efforts and demanding subsistence practices; however, it is also explicitly and emphatically gendered. We depart from this gendered perspective ideology, and suggest an alternative discourse that is inclusive of both women and men: humanism. Modern development attitudes exploit not only women and nature, but all human beings. Therefore, there is a need for a humanism that transcends the parochial view of gender relations and offers a universalist perspective.

The Sundarbans: A Conversation between Humans and Non-Humans

Gender and the environment are relatively young fields of academic study, and are interwoven with the theme of gender politics, environmental crisis and development. But how exactly are gender and environment connected in these fields? Answering this pressing question is a necessary preliminary to defining the practical relationship between gender and environment in the Indian context and beyond. Environmental studies began in geology, biology and meteorology, before gradually widening its periphery to include the humanities and social sciences. The environment includes not just flora and fauna but all living beings, including human beings. Moreover, the concept of gender is complex and "there are numerous forms and expressions of gender, of 'being masculine' and 'being feminine'" (Beynon 2002, 1). Often, the concept of gender creates dichotomy, but if we probe into the discursive formation of this concept, it is obvious that

> an essentialist "male–female" dichotomy cannot account for the ways people are gendered in different places at different times [...] the idea of "being a man" can no longer be treated as fixed or universal [...] Masculinity is inclusive of male and female, and equally femininity is inclusive of female and male.
>
> (Cornwall and Lindisfarne 2003, 3)

By this we infer that gender is fluid and a cultural borrowing, and "it would appear that we live at a time when gender identity has less to do with biology than with economic and cultural circumstances" (Beynon 2002, 6). In our view, gender shapes interhuman and non-human relations, and the environment shapes human relations. The concept of gender thus not only relates to masculinity and femininity, as discussed by Cornwall and Lindisfarne, but also embraces the realm of human and non-human interaction, as our analysis of Ghosh's *The Hungry Tide* shows.

The Hungry Tide is set in the Sundarbans, which is also known as the tide country. The novel is an amalgamation of the real history of the Sundarbans and fictional elements. It is about interdependency, interrelation and interaction among humans, forests, water and non-human beings. This symbiotic relationship of humans and non-humans that Ghosh presents can be traced back to India's ancient past, the Vedic period, in which human settlements were generally found near lakes, rivers, hills and forests, which people depended on for their livelihood; settlements were especially common in the Gangetic Plain prior to its deforestation. The main occupation in this era was farming and cattle rearing. Human societies were primarily pastoral and based on possession of cattle and horses, which were synonymous with wealth. These societies were characterized by a harmonious relationship between humans and non-humans (Sharma 2005). Today, 70 per cent of the Indian population still live in villages and their main occupation is farming. The majority of Indians thus dwell in rural areas and are dependent on nature. Ghosh interweaves this human and non-human interdependency with two temporal narratives. The first is Nirmal's narration of the Marichjhapi massacre in his journal[1], which took place 28 years before the time the novel is set; the second is Piya's expedition, which explores the contemporary situation of the Sundarbans with a particular focus on the ecosystem. These two narratives, the past and the present, are of central importance for comprehending the dialogue between humans and non-humans. The detailed description of the Sundarbans' flora and fauna is an interstitial space for humans and non-humans to interact. It is a topophilia[2] that maps the cognitive, emotional and physical bonds that exist between the people, the land and animals of the Sundarbans. The dominant mode of narration is realism, as developed in the tradition of the Indian novel. The narrative revolves around the Sundarbans mangrove forest. The Bengali name Sundarbans is etymologically derived from *sundar*, meaning beautiful, and *ban*, meaning forest; that is, a beautiful forest. It is also derived from the *sundari* trees that are abundant in this mangrove forest. The Sundarbans covers an area of 10,000 square kilometres, stretching across the Indo-Gangetic Delta in West Bengal, India. It is the largest mangrove forest and tiger habitat on the globe, and was declared a wildlife sanctuary in 1977 and a national park in 1984. The tigers in this area are protected under the Tiger Conservation Project launched in 1973 and it is a UNESCO World Heritage Site. According to the Indian government website and the Geological Survey of India, it was declared a reserved forest in 1875–1876 under the 1865 Forest Act.

The novel offers a spatial history of the Sundarbans, an archipelago stretching from Hooghly River in West Bengal to the Meghna River in Bangladesh. It is made up of thousands of islands, some larger, others more like sandbars. The river channels spread across the land,

Yet each of these channels is a "river" in its own right, each possessed of its own strangely evocative name. When these channels meet, it is often in clusters of four, five or even six: at these confluences, the water stretches to the far edges of the landscape and the forest dwindles into a distant rumour of land, echoing back from the horizon (Ghosh 2008, 7).

There is a continuous convergence between forest and tides; everyday tides engulf thousands of acres of forest, which re-emerge after a few hours – an ecological saga. The peninsulas and promontories change their shapes due to powerful currents, and sometimes the islands vanish and new islands emerge. It is a game of hide and seek between land and water: "When the tides create new land, overnight mangroves begin to gestate, and if the conditions are right they can spread so fast as to cover a new island within a few short years" (8). The mangrove forest of the Sundarbans is a universe in its own right. Mangrove leaves are tough and leathery; the forest is a very dense habitat of tigers, crocodiles, snakes, dolphins and so on. The novel very clearly projects the attitude of human beings towards nature: "At no moment can human beings have any doubt of the terrain's utter hostility to their presence, of its cunning and resourcefulness, of its determination to destroy or expel them" (ibid.). Usually, humans are characterized as dominating the non-human world, as, for example, in the narrative of the Anthropocene. However, the novel inverts this cultural construct and shows that the tides dominate humans. On the one hand, the tide constantly sweeps over the land and people's lives, while on the other, animals encroach on the human settlement and take a toll on their lives.

The land is always half-submerged because of high tides, and the water gives birth to the forest. The Mughal emperors named this region *bhati* – tide, or rather not just tide but ebb tide, *bhata*, and the land is called *bhatir desh* – tide country:

> A study [...] had shown that there were more species of fish in the Sundarbans than could be found in the whole continent of Europe. This proliferation of aquatic life was thought to be the result of the unusually

> varied composition of the water itself. The waters of river and sea did not intermingle evenly in this part of the delta; rather, they interpenetrated each other, creating hundreds of different ecological niches, with streams of fresh water running along with the floors of some channels, creating variations of salinity and turbidity.
>
> (Ghosh, 131)

This detailed description of the Sundarbans shows that this landscape is not a mere ecosystem but a vibrant assemblage and presence of non-humans. It is both a real and an emotional space that the novel's diverse characters enter for various reasons; through their interactions, the novel explores the interpolation of humans, culture and ecosystem. All the characters have an encounter with the tidal land and their destiny is dictated by natural and sociocultural geopolitics (something which is both implicitly and explicitly discussed), by wide-ranging relations of inequality, materiality, power, violence and colonization, by disparities at various levels and by gender and the environment.

The novel narrates the colonial past through the dreams of Daniel Hamilton, a rich Scotsman, and the postcolonial present through the characters Piya Roy and Kanai Dutt. Driven by Daniel's dominating attitude, human beings enter the tide country. The place is named Lusibari, derived from the English name "Lusy's house". It is a long way upriver. Gradually, people start pouring into the land, intent on making a new home, and begin killing tigers, crocodiles and snakes to keep the place safe for themselves. However the land resists their attempts to tame it. They cannot grow anything, because salt and tides continuously flood the fields. People therefore live in utter poverty and hunger, and Catastrophe becomes their way of life, but they never complain. Instead, they learn to live in adverse conditions. This is a dialogue between humans and nature in a form of adversity. Nirmal, a Marxist intellectual and writer who teaches at a Lusibari school, writes in his journal about this tidal land:

> Now, with the rising of the sun, I have understood what it is: I am afraid because I know that after the storm passes, the events that have preceded its coming will be forgotten. No one knows better than I how skillful the tide country is in silting over its past.
>
> (Ghosh, 74)

The storm is an allegory of nature's superiority over the human. It takes away people and even animals. But animals know how to commune with nature. The silent activism of nature has taken steps to ensure its survival by sweeping away the most adamant and stubborn attitudes that humans espouse in the name of progress.

In another episode, Piya Roy, a cetologist of Indian origin based in the United States, visits Canning to do a survey of the marine mammals of the Sundarbans, in particular the Orcella dolphin. While conducting the survey with her government guide and guard, she accidentally falls into the river; a young and illiterate fisherman, Fokir from Lusibari, rescues her. The two are able to communicate extensively without language, and a strong bond grows between them. Fokir helps Piya in her survey by taking her in his boat, but nature's fury, in the form of a flood, takes its toll on him. Kanai, a businessperson from Delhi, comes to Lusibari at the request of his aunt Nilima to read her husband Nirmal's memoirs, recorded in a journal that he left behind. The novel revolves around these characters from different worlds, who come to Lusibari for different purposes, and thereby negotiates a wide range of social relations: between the Global North and the Global South, between urban and rural experiences, between different classes, culture, caste, genders and ages:

> A growing body of work shows how people's interactions with animals are not only embedded in their relation to the environment as a practical experience but also as narratives, giving them scope to elaborate on their ideas of the social.
>
> (Jalais 2010, 7–8)

The character of Piya and her relation to Fokir are particularly interesting in this context. In the Sundarbans, the tide country, transformation is the rule of life:

> Rivers stray from week to week, and islands are made and unmade in days. In other places forests take centuries, even millennia, to regenerate; but mangroves can recolonize a denuded island in ten to fifteen years. Could it be that the very rhythms of the earth were quickened here so that they unfolded at an accelerated pace?
>
> (Ghosh 2008, 241–242)

It also transforms the postcolonial outlook of Piya and Kanai – Organic Intellectuals. Piya's core motive is to do research on marine mammals (dolphins) – an elite ecological approach. For her, the Sundarbans is a natural ecosystem, a place for research, and after her project she will return home, whereas Fokir, who lives there, depends on the Sundarbans for his being and existence – he is a traditional intellectual. Fokir's intimate knowledge of and association with nature is based on his experience of dependency on, and strong bond with, nature. Over the course of time, Piya, a representative of Eurocentric modern knowledge, acquires not only academic knowledge but also some of Fokir's native expertise regarding the ecosystem:

> It is a connection that springs from their common interest in and knowledge of dolphins, waterways, and nature, a connection that is in some way immediate, but that, for her at least, could only be reached through her immersion in the world of mediated knowledge of this place.
>
> (Weik von Mossner 2007, 127)

Piya's project and perspective shift from a merely scientific endeavour to a search for experiential knowledge of the ecosystem, by which she becomes a part of it. Later, she settles in Lusibari to help the local fishermen and Fokir's family. Kanai works as a transnational entrepreneur, representing global capitalism. Initially, he thinks of the Sundarbans simply as a dense forest. However, his continuous interaction with Nirmal's journal, which tells the story of the island, and his expedition with Piya and Fokir brings about profound changes in him. The characters are bound to particular places, and "position themselves in natural settings in order to reinhabit a landscape or place that is intrinsic to their philosophies of being in the world" (Dreese 2002, 6). Intruders like Daniel Hamilton, Kanai and Piya impose their own culture, which is a marker of civilization and the Western development away from an eco-based way of life.

The analysis of the novel reveals how the concern with ecology depends on space hence on the geopolitics and socioculture of particular heterotrophs. Lusibari exemplifies the people who came to live on this island and try to tame and control the ecosystem for their basic needs. Gandhi very aptly said, "Nature has enough for need, not for our greed". We can see this harmonious and reciprocal relationship of humans and nature in certain examples. There are more than 500 tribes in India who can be considered true environmentalists and serve as examples for a harmonious and reciprocal relationship between humans and nature. The Bishnoi tribe were the first environmentalists and eco-conservationists, and through the Chipko movement they earned worldwide fame. They have 29 religious principles, six of which are dedicated to the protection of the environment, including *jeev daya paloni* (be compassionate to all living beings) and *rukh lila nahi gahave* (do not cut green trees). In present-day India, there are still many tribes who dwell in the forest region and have an ecological sensibility, and remain untouched by economic and technological progress. Moreover, for them, nature is not just a means of sustenance, but represents their rootedness and the reproduction of their society. Though the Chipko movement was led by women, men also participated and played a vital role in the non-violent protest against the felling of the trees, including Gaura Devi, Bachni Devi, Chandi Prasad Bhatt, Sunderlal Bahuguna, Govind Singh Rawat, Doom Singh Negi and Ghanasyam Raturi. As Sunderlal Bahuguna, a leading figure in the Chipko movement, crusading activist and environmentalist, explains,

> But this time, they said, we'll hug the trees. So, the idea came, they demonstrated and the company people ran away from there. People went from village to village with this idea. In Indian history, in 1730, as many as 363 men and women offered themselves to the axe, in a village near Jodhpur in Khejadli village and they were the Bishnoi people. The tradition among the Bishnois is that they won't allow anybody to fell trees and kill wild animals. That was history though we didn't know about it then.
> (Bahuguna 1997, 15)

Other environmental movements and protests, such as the Appiko movement (Gadgil and Guha 1994), the Save the Narmada River movement (Rao 2012) and Save Silent Valley (Chengappa 2009), also had equal involvement of men and women. Keeping these examples in mind, the spatial narration of Lusibari describes an affective bond between people and their natural environment. Vandana Shiva aptly notes that

> In tribal and peasant societies, cultural and religious identity derives from the soil, which is perceived not as a mere "factor of production" but as the very soul of society. The soil has embodied the ecological and spiritual home for most cultures. It is the womb not only for the reproduction of biological life but also for cultural and spiritual life; it epitomizes all the sources of sustenance and is "home" in the deepest sense.
>
> (Mies and Shiva 2010, 102)

In order to present a reasoned argument, we use a comparative perspective to support certain lines of argument and also provide counterpoint aspects of Indian ecofeminists' views: Within her critique of development, progress and globalization, Vandana Shiva has subsumed all third-world women under a single category instead of understanding women's identities more heterogeneously and based on an intersection between caste, class, race and culture. She ignores the fact that power, privilege, property and caste/class stratification predate colonialism and have become the principle causes of inequalities and destruction of nature (Agrawal 1992, 119). Chhaya Datar (2011), an Indian ecofeminist thinker, claims local production system and subsistence practices of rural women, and Aruna Gnanadason (2003) opposes economic liberalization and privatization of every sector which impacts on livelihood of villagers. However, theses perspectives have specific limitations and require a substantial departure and so propose a new perspective – ecohumanism.

To this end, we refer to *The Hungry Tide* which helps us to develop a more complex understanding of gender and environment relations. First, the detailed discussion of the flora and fauna of the Sundarbans provides a space to meditate on the interrelationship between humans and non-humans. Second, Lusibari, the first human settlement in the tide country, which brings with it the dominating Anthropocene attitude, exemplifies a compatible and convergent relationship between humans and non-humans.

Human and Non-Human Dynamics

The novel is divided into two sections: the ebb (*bhata*) and the flood (*jower*) constitute a fluid space where the interplay of social, spiritual and ecological identities is constructed. The story revolves around the forest and wild animals, principally crocodiles, tigers, dolphins and snakes, and the dangerous tides which sweep through people's lives – human and non-human dynamics. For the characters in Ghosh's novel, nature is not only a means of survival, but something with which they have a deep-rooted spiritual connection. This resonates with popular culture, in which many people worship the sun, moon, mountains, rivers, trees, animals and plants (and have done so since the Vedic period) because they regard them as living beings with human attributes. Many trees and plants such as vat, people, neem, tulsi and shami are worshiped and considered sacred because of their medicinal properties (Sharma 2005). This reverential attitude and the connection between humans and non-humans are portrayed in the novel through the story of the Ganga and Bon Bibi. The novel draws on the different mythical narrations of the origin of the Ganga river and the divine and holy connections between humans and nature, and also discusses gender and nature relations with respect to Indian religious traditions. To give a brief summary: the Ganga river is considered a goddess. She descended from the abode of God. Lord Shiva tamed Ganga in his knotted lock of hair. When he undid his braid, the river flowed from thousands of tangled hair strands onto the earth: "Once past that point the river throws off its bindings and separates into hundreds, maybe thousands, of tangled strands" (Ghosh 2008, 6). The other story is that one time, Lord Vishnu, in the guise of a dwarf, began the task of measuring the earth with his giant strides, and in one place made a small hole with his toe from which the Ganga started flowing onto the earth to wash away people's sins.

The novel also assembles the polemics and spiritual discourses from the archaic folk story of Bon Bibi and Shah Jongoli, which reflects on the interconnection of humans and non-humans. These twins have been given the task of making the jungle of the country of 18 tides, the Sundarbans, fit for human settlement. The land was previously the domain of the demon Dokkhin Rai, who hates human beings and craves their flesh. There is a battle between Dokkhin Rai and Bon Bibi. Despite emerging victorious, Bon Bibi mercifully decided that

> one half of the tide country would remain a wilderness; this part of the forest she left to Dokkhin Rai and his demon hordes. The rest she claimed for herself, and under her rule once-forested domain was soon made safe for human settlement. Thus, order was brought to the land of eighteen tides, with its two halves, the wild and sown, being held in careful balance. All was well until human greed intruded to upset this order.
>
> (Ghosh, 108–109)

This story explains the equilibrium of humans and non-humans, and intimate ties to a specific environment: "The forest fishers enter the forest under the protection of the forest goddess Bon Bibi and see their line of work as 'respecting' the forest as it neither 'depletes' nor 'disturbs' it" (Jalais 2010, 31). These two mythological stories account for our deep-rooted spiritual values and reverence towards nature. The local people have a close connection with the non-humans:

> Piya, Kanai or Nirmal – products of the cosmopolitan process – narrate the panoramic view of the Sundarbans and discuss its marine or human ecosystem whereas Fokir, Moyna, Horen or Tutul are part of the landscape: they do not narrate the place, they live the landscape.
>
> (Rath and Malshe 2010, 29)

Being a fisherman, Fokir has a deep knowledge of rivers, forests and animals. He is very close to nature and has an eco-friendly attitude. His fishnet is very strange, not the usual nylon net: "Stranger still, there were no hooks. Instead, spaced between the weights, were bits of fish-bone and dried cartilage, tied to the line with a cord" (Ghosh 2008, 147). Each aspect of his life is like a grassroots ecologist, and his lifestyle is harmonious with nature:

> Through Fokir's character, the novel seems to inquire: who is actually an outsider in the ecosystem of tidal land? Are the "native" fishermen responsible for destroying the natural ecosystem or are the forces of industrialization, urbanization, and forms of government and semi-government interference responsible for the actual destruction of the ecological balance?
>
> (Rath and Malshe 2010, 24)

These native people are never anti-ecology. They are dependent on forests and rivers, and so there is a harmonious relationship between them. The mutual understanding and the human–animal dynamics are exhibited through an astonishing story: Piya and two fishermen go to survey the river. After they have covered some distance, the older fisherman begins to beat the drum on the gunwale of the boat. A few dorsal fins come up to the surface; the younger fisherman takes the fishnet and makes a rattling sound, which summons a pair of dolphins to the boat. As they approach, they start to move in a circle. "Through the interpreter, the fisherman explained that the dolphins were herding a school of fish towards the boat" (Ghosh 2008, 178). A euphoric synergy between voiceless and voiced:

> Did there exist any more remarkable instance of symbiosis between human beings and a population of wild animals? She could not think of one. There was truly no limit, it seemed, to the cetacean gift for springing surprises.
>
> (Ghosh, 179)

Kusum, Fokir's mother, constantly feels that animals are given higher priority than human beings, as the government is trying to evacuate the land for the sake of tigers. In this pressing situation, she comes to Nilima for medical assistance for the refugees from Marichjhapi. Nilima says, "How can they just seize it? If they're allowed to remain, people will think every island in the tide country can be seized. What will become of the forest, the environment?" (229). She also gives a contrasting view: "They are providing water for tigers! In a place where nobody thinks twice about human beings going thirsty!" (260). Another example of this perspective is Piya's ecological concern, which becomes apparent in her response to two incidents: the killing of a tiger by the villagers, and the death of a dolphin. Despite knowing that the tiger has killed two people and several domestic animals, Piya tries to stop the villagers from killing it: "This is an animal, Kanai. You can't take revenge on an animal" (317). On top of that, a fast-moving motorboat, belonging to officials from the coastguard or the police or the forest department, kills a dolphin that gets caught in the propeller as the boat sweeps ruthlessly through the water. Piya's philosophy of upcoming disaster resonates with Nilima's disquieting worry:

> But by nature, by the earth, by the planet that keeps us all alive. Just suppose we crossed that imaginary line that prevents us from deciding that no other species matter except ourselves. What'll be left then? Aren't we alone enough in the universe? And do you think it'll stop at that?
>
> (Ghosh, 326)

Let us move to another Sundarbans island: Marichjhapi, the home of tigers, a tide country island a couple of hours from Lusibari by boat. It is a part of the Sundarbans forest reserve dedicated to tiger conservation. One notable feature of this island is that although it is easily accessible from the mainland, it is not populated by people. The novel gives a human face to the reality of the Marichjhapi massacre, which took place in 1978, after which several refugees from Bangladesh came to the island to take shelter. Most of them were Dalits. They came from a government resettlement camp in central India:

> In the year after Partition the authorities had removed the refugees to a place called Dandakaranya, deep in the forests of Madhya Pradesh, hundreds of kilometers from Bengal [...] It was more than a concentration camp or prison. They were surrounded by security forces and forbidden to leave. Those who tried to get away were hunted down.
>
> (Ghosh, 124)

Somehow, some of the refugees managed to escape and came to Marichjhapi. Among them were Kusum and Fokir, who left the island long ago.

While Kusum is talking about her journey, she explains the importance of soil and land:

> We'd never seen such a place, such a dry emptiness; the earth was so red it seemed to be stained with blood. For those who lived there, that dust was as good as gold, they loved it just as we love our tide country mud. But no matter how we tried, we couldn't settle there, rivers ran in our heads, the tides were in our blood.
>
> (Ghosh, 174–175)

But in 1979, the island was evacuated, either by sending people back to their country or leaving them to die of disease or malnutrition or in the conflict unleashed by the government. It is one of the worst massacres in the history (Halder 2019). The island is now reserved for tiger conservation, and so human beings are not allowed there. This real-life story is retold in the novel:

> Within a matter of weeks they had cleared the mangroves, built *badhs* and put up huts. It happened so quickly that in the beginning no one even knew who these people were. But in time it came to be learnt that they were refugees, originally from Bangladesh.
>
> (Ghosh 2008, 124)

Initially, they lived a very eco-friendly life. But gradually things changed; with the introduction of modernity, there ensues destruction of nature and natural resources, due to human greed rather than need:

> Saltpans had been created, tube-wells had been planted, water had been dammed for the rearing of fish, a bakery had started up, boat-builders had set up workshops, a pottery had been founded as well as an iron smith's shop; there were people making boats while others were fashioning nets and crab lines; little marketplaces, where all kinds of goods were being sold, had sprung up. All this in the space of a few months! It was an astonishing spectacle – as though an entire civilization had sprouted suddenly in the mud.
>
> (Ghosh, 204)

When the government tries to evict people from the land, we witness the helpless plight to which human beings are subjected for the sake of non-humans, through the heart-rending dialogue of Kusum:

> It was to sit here, helpless, and listen to the policemen making announcements, hearing them say that our lives, our existence, was worth less than dirt and dust. "This island has to be saved for its trees, it has to be saved for its animals, it is a part of a reserve forest, it belongs to a project to save tigers, which is paid for by people from all around the world.
>
> (Ghosh, 284)

The crux of ecohumanism, a reciprocal relation between humans and non-humans, is showcased by Ghosh, first, in an illuminating traditional religious story/myth concerning the sacred role of the river in human beings' life and the decision of the regional god and goddess to protect the wilderness in half the forest while allotting the remaining half to human beings. Second, the Marichjhapi incident shows the importance of non-humans for humans. In the next section, we will focus on the global and local characters of the narrative that support our proposal for an ecohumanist perspective.

Beyond Binaries: Constructions of Gender and Environment

A close reading of Amitav Ghosh's *The Hungry Tides* allows us to flesh out an ecofeminist perspective, based on the interrelationship between humans and non-humans. In particular, our analysis of the Marichjhapi incident and the analysis of various characters, both cosmopolitan and local, contribute to a novel reconceptualization of the relation of gender and environment that goes beyond binaries. The interaction and dynamics of humans and non-humans in the Sundarbans creates a space to revisit the Indian cultural, philosophical and spiritual resources that we use to comprehend the relationship of gender and environment, and in a postcolonial intellectual climate enquires into the validity of the tropes of postcolonial ecofeminism, "'the double bind' of being female and being colonized" (Campbell 2008, xi), and notions of class, caste, race, colonialism and neocolonialism. Non-binary thinking is deeply entrenched in traditional Indian culture. The well-known tenth-century *Bhakti* ("devotion") movement rejected gender binaries, believing that gender is inclusive of both male and female; accordingly, both shift their identity and enact them in the same way as characterized in of environmental narratives. Some saints of the Bhakti movement, such as Akka Mahadevi, Aydakki Lakkamma, Devara Dasimayya, Allam Prabhu and Basaveshwara, believe that "Sharana sati, Lingapati" (Sharana (devotee) is the wife and Shiva (God) is the husband). In this context, it is worth reflecting on H. S. Shivaprakash's[3] claim that "when it comes to *Bhakti,* even the male becomes female with respect to the deity" (2018). In these examples of non-binary thinking, neither man nor woman is a monovocal subject; they are indexical of ecological zones, classes, castes and races, so that neither is homogeneous. Only a false generalization, a form of essentialism, conflates all men or all women in one category. "If gender is cultural, then it follows that women, as well as men, can step into and inhabit (whether permanently or temporarily) masculinity as a 'cultural space', one with its own sets of behaviors" (Beynon 2002, 7). Furthermore, the intrinsic value of nature for human life constitutes – from the Vedic period to the contemporary era – an intimate relationship between ecology and humans. The Indian Vedas worship the earth as the divine mother goddess *Bhumi-Devi.* In the *Atharva Veda*, the earth is worshipped in the "hymn to the earth" (*Bhumi Suktha*). It proposes a holistic relationship between humans and nature, and reveres each constituent of nature, such as *sthalavriksham* (from *sthala* meaning place and *vriksham* meaning tree), which is intimately associated with the deity and the related temple. We worship animals, rivers, mountains, the sun, the moon and so on. Each tree tells a story. The *Vedas* include 1,028 hymns about love of nature (Kermani 2017). The *Taittiriya Upanishad* sets out the five basic elements of life (*panchamahbhuthis*): *akash* (sky/ether), *vayu* (air), *tejas* (fire), *ap* (water) and *prathavi* (earth). All living beings are made up of *panchamahbhuthis*, which means there is an interconnection between humans and non-humans. One of the six ancient Indian philosophies is *Samkhya*, which emphasizes the concepts of *prakriti* (nature) and *purusha* (man), as "the world owes its creation and evolution more to nature or *prakriti* than God" (Sharma 2005, 276). All four Vedas (*Rig*, *Sama*, *Yajur*, *Atharva*) talk about the web of human life, not just with other humans, but with non-humans too. Nature worship still exists today in India in the world of tribal people. This discussion is a backdrop to understanding the sociocultural context of the novel. In the Sundarbans, tides are continuously engulfing the land, forest, animals and people, and two fierce storms have taken a toll on human beings by sweeping away their habitat. There is a fluid exchange of land and water where neither prevails. On the other hand, the Sundarbans is not merely an ecospace, but a fertile ground for examining the historicity of the construction of gender and environment in an Indian context. There is a lot happening in two worlds, nature and human beings, the former symbolizes the Indian ecosystem/environment and latter represent actors as well as victims. To contextualize the discussion, on the one hand, there is the entry of the Anthropocene into the Sundarbans through the clearing of mangroves by government-sponsored agriculture (Jalais 2005). On the other hand, there is a need to reflect on the victims of the Sundarbans, those whose lands, crops and lives have been swept away by the tides, as well as the entry of the tigers into the human settlement. The Sundarbans, Lusibari and Marichjhapi episodes paint a rich picture of the communion of human beings and nature. Humans are both supporters and nurturers of nature as well as victims of its fury. Let us consider a few examples of human concern for nature. Horen represents the native people's attitude to nature; they live in harmony with the natural world. When Nilima asks him why he went to the jungle, he responds, "I went to 'do jungle' [...] and Bon Bibi granted me enough honey to fill two bottles. I came here to sell them" (Ghosh 2008, 29). Bon Bibi, the goddess of the forest, offers him honey, which he accepts and takes to the market to sell so he can make a living. On the other hand, Moyna feels very proud to show all the progress in the town to Kanai, a cosmopolitan influence. "Look at the shops and stalls that have come up around the hospital. See how many there are?" (139). In addition, she dislikes the primitive lifestyle of her husband and the organic community. She is an ambitious woman, a symbol of development, with a progressive way of thinking and opposed to her son for following her husband's profession, "because Tutul has to go to school, doesn't he? [...] I don't want him growing up catching crabs. Where's the future in that?" (140). However, she forgets that Fokir is her contemporary, and not from a bygone age. In real terms, he is a developer. These instances show that men do not always have a paternalistic outlook and women do not always have a feminine quality. "They explain how they respect the forest because they enter it 'with peace', consider tigers to be their 'brothers' and consequently share the products of the forest equitably between wild animals and themselves" (Jalais 2010, 31). Nirmal always thinks that everyone is equally a part of the universe. His ecological outlook is expressed in this ideology: "For him, it meant that everything which existed was interconnected: the trees, the sky, the weather, people, poetry, science, nature" (Ghosh 2008, 306).

Our analysis of the novel in terms of the relation between humans and non-humans has given us a new perspective on ecofeminism: "Ghosh explores a sense of connection between people that transcends class, culture, language, and gender on the basis of a common being and common relations of the natural environment" (Weik von Mossner 2007, 136). Based on these insights, we suggest a shift from ecofeminism to ecohumanism that would allow us to overcome narrow identity politics while acknowledging the existence of multiple inequalities and power relations. An ecohumanist perspective stresses the centrality of living beings, and conceives of human beings as part of the ecosystem. Ecohumanism thus propounds the idea that human beings – both women and men – are part of a complex of nature. We do not claim to have developed this idea into a full-fledged theory, but we would like to propose a working definition according to which ecohumanism "seeks to uphold an equal relation between men and women by dissolving dualistic approaches to resolve the ecological crisis through common consensus". Aldo Leopold presents an analogy that helps to contextualize this discussion:

> To understand the function of a hand it is necessary to understand the whole body and consider the former in an organic relation to the latter. Similarly, a human is both a member as well as the home of which (s)he is a member.
>
> (Leopold 1949, 204)

Meanwhile, Marie Wilson's philosophical understanding adds a nuanced perspective to our ecohumanism:

> Though I agree with ecofeminist analysis, the difference must be because of where I come from. In my mind, when I speak about women, I speak about humanity because there is equality in the Gitksan belief: the human is one species broken into two necessary parts, and they are equal. One is impotent without the other.
>
> (Wilson 1990, 77)

Thus, the relation between gender and environment is not homogeneous; rather, it is a socio-spatial-cultural construction of a particular place/space – a geopolitics.

Conclusion

Within the limited scope of this chapter, we have attempted to articulate the ecologies of gender through an analysis of the novel *The Hungry Tide* and to reconstruct the human–nature dynamics and geographical conditions of the Sundarbans as presented by Ghosh. In our view, the novel can be read as a contribution to ongoing debates on ecofeminism and the non-human turn because it deconstructs binary constructions of gender through an environmental narrative. The novel clearly highlights the harmonious connection of human beings with nature as well as the conflicts among humans and contested understandings of ecology in the tigers versus refugees' episode in Marichjhapi, where people scrape a living from the scanty resources of the forest and from fishing. The novel also historicizes the polemic narratives of the relationship between gender and nature, and answers the questions raised by gender-oriented discourses. To address environmental problems, we need men and women to join together, and so ecofeminism needs to take a back seat and has to be rethought as ecohumanism – as a connectedness and reciprocal understanding between humans and non-humans. This approach stresses the centrality of living beings, not the centrality of human beings.

Notes

1 Nirmal recorded his experiences in his diary that is called journal, which plays a central role in the novel.
2 W. H. Auden used this term for the love of nature, especially love of place. Here, we would like to indicate a close cultural connection to a certain place.
3 A well-known Indian poet and playwright who writes in Kannada and English.

References

Agrawal, Bina. 1992. "The Gender and Environment Debate: Lessons from India." *Feminist Studies* 18 (1): 119–158.

Bahuguna, Sunderlal. 1997. *Walking with the Chipko Message*. Garhwal: Navjeevan Asharam.

Bahuguna, Sunderlal. 2007. *Dharti Ki Pukar*. New Delhi: Radhakrishna Prakashan.

Beynon, John. 2002. *Masculinities and Culture*. Buckingham and Philadelphia: Open University Press.

Bhatnagar, O. P. 2007. *Indian Political Novel in English*. New Delhi: Sarup & Sons.

Campbell, Andrea, ed. 2008. *New Directions in Ecofeminist Literary Criticism*. Cambridge: Cambridge Scholars Publishing.

Chengappa, Raj. 2009. "1976- Silent Valley Movement: The Genesis of Green." *India Today*, December 24. https://www.indiatoday.in/magazine/cover-story/story/20091228-1976-silent-valley-movement-the-genesis-of-green-741606-2009-12-24.

Cornwall, Andrea, and Nancy Lindisfarne. 2003. *Dislocating Masculinity: Comparative Ethnographies*. London: Routledge.

Datar, Chhaya. 2011. *Ecofeminism Revisited*. Jaipur: Rawat Publications.

Dreese, Donelle. 2002. *Ecocriticism: Creating Self and Place in Environmental and American Indian Literatures*. New York: Peter Lang.

Fletcher, Lisa. 2011. "Reading the Postcolonial Island in Amitav Ghosh's *The Hungry Tide*." *Island Studies Journal* 6 (1): 3–16.

Gadgil, Madhav, and Ramachandra Guha. 1994. *Ecology and Equity: Steps towards an Economy of Permanence*. Geneva: UNRISD.

Gadgil, Madhav, and Ramachandra Guha. 1994. "Ecological Conflicts and the Environmental Movement in India." *Development and Change* (25): 101–136.

Ghosh, Amitav. 2008. *The Hungry Tide*. Gurgaon: Penguin Viking.

Ghosh, Amitav. 2015. *Sea of Poppies*. Gurgaon: Penguin Books.

Ghosh, Amitav. 2016. *The Great Derangement: Climate Change and Unthinkable*. India: Penguin.

Gnanadason, Aruna. 2003. "Traditions of Prudence Lost: A Tragic World of Broken Relationships." In *Ecofeminism and Globalization*, edited by Heather Eaton and Lois Ann Lorentzen. New York: Roman & Little Field Publishers.

Halder, Deepa. 2019. *Blood Island: An Oral History of the Marichjhapi Massacre*. Noida: Harper Collins.

Jalais, Annu. 2005. "Dwelling on Morichjhanpi: When Tigers Became 'Citizens', Refugees 'Tiger-Food'." *Economic and Political Weekly* 40 (17): 1757–1762.

Jalais, Annu. 2010. *Forest of Tigers: People, Politics and Environment in the Sundarbans*. Abingdon: Routledge.

Kaushik, Asha. 2001. *Politics, Symbols, and Political Theory: RethinkingGandhi*. New Delhi: Rawat.

Kermani, Vivek. 2017. "What Modern Ecology Can Learn from Ancient Hinduism." *Ecologise*, 19 April 2017. Accessed 1 June 2021. https://ecologise.in/2017/04/19/what-modern-ecology-can-learn-from-ancient-hinduism/.

Leopold, Aldo. 1949. *A Sand County Almanac and Sketches Here and There*. Oxford: Oxford University Press.

Naik, M. K., ed. 1985. *Perspectives on Indian Fiction in English*. New Delhi: Abhinav Publications.

Nair, Varsha. 2018. "Translating Akkamahadevi's Vachanas: In Conversation with H. S. Shivaprakash." Accessed 1 June 2021. https://www.sahapedia.org/translating-akkamahadevis-vachanas-conversation-hs-shivaprakash.

Rao, Manisha. 2012. "Ecofeminism at the Crossroads in India: A Review". *DEP* (20): 124–142.

Rath, Arnapurna, and Milind Malshe. 2010. "Chronotopes of 'Places' and 'Non-Places': Ecopoetics of Amitav Ghosh's *The Hungry Tides*." *Asiatic* 4 (2): 14–32.

Sharma, R. S. 2005. *India's Ancient Past*. New Delhi: Oxford University Press.

Shiva, Vandana. 2010. *Staying Alive: Women, Ecology and Survival in India*. New Delhi: Women Unlimited.

Sultana, Farhana. 2020. "Embodied Intersectionalities of Urban Citizenship: Water, Infrastructure, and Gender in the Global South." *Annals of the American Association of Geographies* 110 (5): 1407–1424.

Thompson, Jennifer. 2016. "Intersectionality and Water: How Social Relations Intersect with Ecological Differences." *Gender, Place & Culture* 23 (19): 1286–1301.

Weik von Mossner, Alexa. 2007. "The Home, the Tide, and the World: Eco-Cosmopolitan Encounters in Amitav Ghosh's *The Hungry Tide*." *Journal of Commonwealth and Postcolonial Studies* 13 (2): 120–141.

Wilson, Marie. 1990. "Wings of the Eagle." In *Turtle Talk: Voices for a Sustainable Future*, edited by Christopher Plant and Judith Plant. Philadelphia: New Society Publishers.

Part 4

Temporalities

Histories, Presents and Futures to Change

11 The Figure of the Human

Philosophical Narratives on Sex, Race and Organic Kinship in the "White (M)anthropocene"

Susanne Lettow

According to the narrative of the Anthropocene, humanity has entered a new geological epoch in which humans have become actors capable of transforming the geological state of the earth. As controversies about the periodization of this new epoch make clear, the concept of the Anthropocene involves a complex array of ethico-political perspectives. Depending who you ask, the new epoch started with the Industrial Revolution, with the "Great Acceleration" of industrialism after the Second World War, with the rise of European capitalism in the wake of the colonization of the Americas or even when humans first became sedentary. Obviously, these different diagnoses correlate with different understandings of what needs to be changed in order to avoid further destruction of the living conditions of humans, animals and plants. Because it is highly disputed what subjects and what kind of societal relations and practices are covered by the term "Anthropos", a wide range of new concepts have been formulated that are intended to render explicit the relations of power and domination, exploitation and extractivism that shape the present epoch. Concepts such as *Capitalocene*, *Plantationocene*, *Chthulucene* and *(M)anthropocene* all intervene in the hegemonic narrative of the Anthropocene and give it a different twist.[1] Although they share a critique of the false universalism that the figure of the human promotes, these concepts highlight different aspects of the problem. As the concept of the "White (M)Anthropocene" makes clear, politics and narratives of sex and race are a constitutive dimension of destructive nature relations. This is ignored within the hegemonic versions of the Anthropocene discourse. As Giovanna Di Chiro, who introduced this concept, puts it, the expression "Anthropocene" suggests "that the Anthropos at the root of the term has no sex, gender, class or race" (2017, 489). However, those who seek to define the nature of the present ecological crisis, as well as adequate strategies to solve it, "are, almost exclusively, white, male, academic, and Euro-American" (490). The figure of the human on which the narrative of the Anthropocene builds thus appears to be neutral while it is, at the same time, shaped by a series of interrelated exclusions and power relations.

DOI: 10.4324/9781003023319-16

In this chapter, I want to explore this from the perspective of the history of philosophy. In particular, I think that an analysis of the emergence of the conceptual figure of the human in the late eighteenth century and of the various rearticulations it has undergone since then can shed light on some political, ethical and epistemic problems that still haunt the *nonhuman turn*. Starting from Donna Haraway's claim that "we must change this story, the story must change" (2016, 40) in order to open up new perspectives on and possibilities for sympoietic nature relations, I will show that current criticisms of human exceptionalism remain bound to a large extent to the "old story" of philosophical anthropology. I argue, however, that a sound critique of contemporary extractivist gender and nature relations requires us to scrutinize this story.

In the first part of this chapter, I will analyse the emergence of the figure of the human around 1800. I fully agree with Kay Anderson and Colin Perrin that "the history of the idea that humans are more than merely human animals" has been "largely overlooked" (Anderson and Perrin 2018, 447) in criticisms of human exceptionalism. The focus on the "Western" tradition in general and more specifically on Descartes' distinction between body and mind in the environmental humanities – including ecofeminist thought – has led to a problematic neglect of the tradition of philosophical anthropology. This emphatically anti-Cartesian way of thinking is best characterized by "the claim that the unique mental capacities of humans are owed not to some metaphysical quality of mind or soul, but rather to the distinctiveness of human anatomy" (452). Moreover, as I will show with reference to writings by naturalists and philosophers in the period, the conceptual figure of the human has been shaped since its inception by (proto-)biological ideas of organic kinship linked to hierarchical constructions of sex and race. In the second part of the chapter, I will outline how the figure of the human and associated philosophical narratives were rearticulated in different approaches within philosophical anthropology in the first decades of the twentieth century. The German philosophers Max Scheler, Helmuth Plessner and Arnold Gehlen, who sought to establish philosophical anthropology as a new philosophical subdiscipline, explicitly referred to the anti-Cartesian tradition of the late eighteenth and nineteenth centuries but built on more recent approaches in biology, in particular Darwinism and ethology. Again, these authors conceived of human exceptionalism as attributable to the specific biological character of humans, and of (male) humans as occupying the highest position within a hierarchical order of organic kinship.

In the third part, I will discuss the extent to which criticisms of human exceptionalism in the context of the *nonhuman turn* build on or depart from the history of philosophical anthropology. My discussion will focus on Bruno Latour's distinction between humans and "Earthbound", and on Rosi Braidotti's idea of a *zoé*-centric egalitarianism. Both authors clearly distance themselves from the false universalism of "humanism", and in particular from the racist and sexist articulations that have shaped the figure of the human. Their attempts to overcome the "humanist" or "anthropocentric" intellectual legacies, however, seem to be half-hearted. In light of the central role that the interrelated notions of organic kinship, sex and race have played in the formation and circulation of the figure of the human, relegating "humans" back to their relation to the earth or to life as such does not seem a critical enough departure from that troubling history. This is not only the case because these theoretical operations ignore crucial aspects of the intellectual history of human exceptionalism; these geo- and *zoé*-oriented criticisms of "humanism" also lead to politically and ethically flawed critiques of universalism. From the perspective of a critical analysis of philosophical anthropology, what is problematic is not the utopian perspective that societal nature relations and associated modes of living might indeed become universalizable and not rely on domination and privilege. Much more troubling, in contrast, is the fact that this perspective has been constantly undermined through biological or proto-biological theories of sex and race that have provided justifications for structural inequalities among humans and shaped "the human" as a figure of mastery and domination. In light of this analysis of subverted universalism, I will conclude by discussing how the understanding of "humanity" can be rearticulated in light of Haraway's notion of a "non-generic" or "critical" humanity.

Sex, Race and Organic Kinship: The Emergence of the Human around 1800

According to Michel Foucault, the figure of the human emerged in the late eighteenth and early nineteenth centuries as an "empirico-transcendental doublet" (1994, 318). On the one hand, the human was defined by reference to scientific (especially biological) knowledge, but on the other was regarded as the condition of possibility of any knowledge, i.e. as a transcendental subject. Insofar as the latter is concerned, the figure was constitutive of modern universalism and its political claims of egalitarianism. However, references to its "empirical" underside have constantly undermined these claims, as they were meant to reveal biological differences and hierarchies within the human species. In addition, the human was inthronized as the supreme living being, defined by its capacity to intervene in and transform the nonhuman world not only through its transcendental position but also through the "empirical" knowledges that shaped the figure. It thus needs to be scrutinized first and foremost with respect to the scientific knowledges that have been constitutive of the discourse of philosophical anthropology since the nineteenth century.

When Carl von Linné published his *Systema naturae* in 1735, he introduced the concept of *homo sapiens* and conceived of humans as animals which could be studied and classified in all their "varieties" just like other animals and plants. This laid the ground for later conceptualizations of humans as a biological species, and although philosophers and naturalists, especially Buffon, soon started to question Linné's taxonomic approach, they built on this idea. The question that occupied many philosophers and naturalists in the late eighteenth century was what kind of "real" qualities – in contrast to purely conceptual distinctions – separate organic from inorganic nature. In these debates, growth, regeneration and reproduction came to be understood as central features of life and thus formed the basis for a new material understanding of the unity of organic nature and of kinship between all living beings. Immanuel Kant, for example, who in his *Critique of Judgement* (1790) rejected the idea of a "science of the living" due to his mechanistic understanding of science, nevertheless mused about the possibility of "an actual kinship" among animal species, reaching

> from that in which the principle of ends seems best authenticated, namely from man, back to the polyp, and from this back even to mosses and lichens, and finally to the lowest perceivable stage of nature [...] to crude matter.
>
> (Kant 2008, §80, 247)[2]

His reflections on "organized nature", which were widely read and contributed much to the development of biology as a distinct field of knowledge, show that he understood kinship among living beings as a hierarchical structure in which beings occupy "higher" or "lower" positions. At the same time, Kant also conceived of hierarchies within the human species. In the period between 1775 and 1788, he published a series of essays in which he gave a definition of the concept of race that allowed him to combine a monogenetic and universal understanding of the human species with an account of structural hierarchies within "humanity". According to Kant's definition, the concept of race refers to "hereditary differences" that "persistently preserve themselves [...] over prolonged generations" and are always transmitting half-breeds (*halbschlächtig*) (Kant 2007, 85). It thus refers to a specific form of difference within the human species that is transmitted through reproduction and processes of heredity. This proto-biological genealogical definition of "race" has played a pivotal role in subsequent enquiries, the modern discourse of race and associated biopolitics.[3]

Although Kant's understanding of gender differences was, to be sure, as problematic as his Eurocentric concept of race, he did not treat these differences in terms of natural history or biology. Other naturalists and philosophers of the period, however, started to do so, and conceptualized gender differences with reference to the early life sciences (Honegger 1991; Schiebinger 1991). Towards the end of the eighteenth century, an increasing number of scientists formulated accounts of gender difference by building on comparative anatomy and theories of reproduction. The theoretical account of the German naturalist Carl Friedrich Kielmeyer is a good example of how the hierarchical understanding of organic kinship included a certain idea of sex, even if Kielmeyer himself was not very explicit about it and refrained from excessive ethico-political claims.[4] He formulated what he called a "compensation model of forces", according to which all organisms are structured by the same forces; however, these forces are arranged in different proportions in different species. According to this model, difference and hierarchy are articulated through the idea of increasing complexity. In the "lowest" or least complex living beings, Kielmeyer argues, the force of reproduction prevails, whereas it decreases in the more complex forms that have a higher proportion of the forces of sensibility, irritability, secretion and propulsion. In particular, he claimed that whereas in the less complex animals in which reproduction and unlimited growth prevail "both sexes become more similar, or disappear altogether" (Kielmeyer 2021a, 40), the more complex organisms display a clearer form of difference, so that in the human species sex differences are mostly developed and stable. Accordingly, the human appears as an organism in which the sexes are most clearly distinguished from each other and thus as the most complex being, though one which, nonetheless, "recapitulates" all the lower stages of organic nature in the process of individual becoming. Thus, although Kielmeyer claimed that all living beings are related to each other, he conceived of the "kinship of species" (1938, 123) as a hierarchical system in which the human occupies the highest position. The more complex the organism, he argued, "the bigger becomes the gap by which it is separated from those who are closest and most similar", so that "humankind, as the last link, stands so far removed from all the others" (2021b, 66).

This connection between human exceptionalism and sex differences came to be rearticulated in various ways, which had a profound impact on the cultural process of the "polarization of gender characters" (Hausen 1976) and on the emergence of modern notions of biological sex. Authors attributed masculinity and femininity to the "ranks" and "stages" of the organic world in different and even contradictory ways, though they all supported patriarchal views. The naturalist and *Naturphilosoph* Lorenz Oken, for example, distinguished between three stages of organic life: infusoria, plants and animals. In his view the infusoria, which represent the lowest stage of life, were "male" because they reproduce autonomously, while plants, the second stage, are "female" and "dependent" because their generation depends on the "male" capacity to reproduce. A plant, Oken argued, is discontent with itself and "seeks outside itself a complement on which it could sustain itself" (2007, 145). For this reason, in his view, diremption emerges at the plant stage and hence "the main character of the plant is sex" (145).[5] At the third organic stage – the stage of the animal – diremption remains essential, although the male capacity to reproduce again prevails. For Oken, this gendered hierarchy of developmental stages coincides with human exceptionalism, so that he finally explicitly claimed the human to be, in fact, predominantly "male". Based on his assumption that "the human, the highest of all animals, unites in himself everything noble and dexterous that is distributed among the rest of the living beings" (3), at the end of the book Oken concludes that "the total human species is nothing but a continually propagating male" (195).

I refer to these examples from the early life sciences here in order to show that the figure of the human as a biological species was, from its inception, entangled with ideas of human exceptionalism, sex and race, and thus conceived as a figure of mastery. However, it was not just the life sciences that played a crucial role; the earth sciences also contributed to the emergence of the figure of the human as an "empirico-transcendental doublet".[6] Henrik Steffens, a naturalist and adherent of Schelling's philosophy of nature, for example, formulated a theoretical account of race in his two-volume *Anthropology* (1822) that was based on an understanding of the earth as an organism, i.e. a hierarchical structure which consists of various elements. "As in the animal structure," he wrote, where

> all organs mutually account for each other and are there at one time, although non-developed, all at once representing themselves and together, in an embryonic stage, several points of the earth have been capable of reaching the highest point [i.e. bringing forth humans] *according to their measure*.
>
> (Steffens 1822, 371, italics in original)

The earthly "organs" or "races" were, in Steffens' view, shaped according to their specific place within the system of the earth, which thus appears to be the basis of racial differences and hierarchies.[7]

Steffens also referred to the ancient idea of the female earth, or the "fecund embrace of Uranos and Gaia" (72). He was certainly not the only one to evoke the figure of the female earth. In the mid-eighteenth century, Julien Offray de La Mettrie had mused in his *Système de L'Epicure* (1750) that in its earliest history "the earth must have served as a uterus to man; it must have opened its womb to the human germs, already prepared, so that given certain laws, this superb animal could hatch out" (La Mettrie, cited in Bowler 2015, 89). Kielmeyer later conceived of the earth as the first material cause of what he called the "economy of the organic world", i.e. the systematic order of life, and referred to organisms as the "*original* children of the life-giving [*gebährende*] earth" (2021b, 67, italics in original). Oken, on his part, referred to the earth as "the first organ of the universe" (1805, 162) and to the "eldest cosmogony" according to which "all living beings have emerged from the earth" (163). To conclude: the figure of the human that emerged on the scene of the European humanities, as Foucault has argued, certainly did not suffer from a lack of biological kinship nor from a separation from the earth it inhabited. On the contrary, the "transcendental" dimension of this figure barely concealed the ideologies of sex and race on which it relied, and which were formulated with reference to the most innovative scientific knowledges of the day.

Rearticulating Human Exceptionalism: Philosophical Anthropology, Race and Gender in Early Twentieth-Century Germany

I now turn to another conceptual configuration, namely, philosophical anthropology as it emerged in the first decades of the twentieth century in German academia. Despite their different approaches, Max Scheler, Helmuth Plessner and Arnold Gehlen all attempted to establish anthropology as a philosophical subdiscipline, building on the discourse of anthropology that had emerged at the start of the nineteenth century and reshaping its central epistemic object, the human.[8] This figure still functioned as an "empirical-transcendental doublet". That means that in these accounts, too, the human was conceived as the central object of empirical knowledge and, at the same time, the condition of possibility of that knowledge, i.e. as a subject with distinctive intellectual capacities. Like in the philosophical-anthropological narratives of the late eighteenth and early nineteenth centuries, Scheler, Plessner and Gehlen mainly built on biological knowledge, which, however, had developed new theoretical subfields in the meantime. So instead of general reflections on the specificity of organic nature, theories of reproduction, comparative anatomy and the earth sciences, the philosophical-anthropological theories of the early twentieth century drew mainly on evolutionary biology, primatology, ethology and – in the case of Plessner – Jakob von Uexküll's *Theoretical Biology*.

In line with early nineteenth-century philosophical anthropology, Scheler, Plessner and Gehlen formulated stage models according to which the human was ranked as the supreme being in the organic world. Their theories, however, differed in terms of how they argued for and legitimized human exceptionalism. In his seminal book *The Position of the Human in the Cosmos* (1928), Scheler presented a thorough comparison between plants, animals and humans. In his earlier treatise *On the Idea of the Human* (1913), Scheler, who had converted to Catholicism, had defined the human as the "god-seeker" but in 1928 he abandoned this definition and attempted to discern different stages of the "biopsychical world" (1976, 12), namely, plants, animals and humans. His main goal was still to conceptualize the "exceptional position" (*Sonderstellung*) of the human but he now built his argument on the difference between chimpanzees and humans. In particular, he contested the theory of ethologist Wolfgang Köhler, who, based on his experiments, regarded chimpanzees as intelligent beings. Scheler rejected Köhler's rapprochement of humans and apes based on his distinction between intelligence and spirit, arguing that the latter is a distinctively human feature that enables the human to "sublimate" the "energy of his drives" (*Triebenergie*) and to transform it into "spiritual activity" (45). Scheler transposed the hierarchical structure of the organic world to his account of subjectivity by arguing that the subject needs to constitute an inner relation of self-domination, i.e. to control its drives and desires. In Scheler's view, the human "spirit" is an "assemblage of acts" (42) that has a "monarchic" (51) structure, so that the "higher" acts govern and dominate the "lower" ones.

Whereas Scheler ultimately re-established the dualistic structure of nature and spirit, his contemporary Helmuth Plessner sought to overcome dualism in what he called "philosophical biology and anthropology" (2003, 9). In his *The Stages of the Organic and the Human* (1928), Plessner, who was also a biologist by training, mainly referred to the biologist Jakob von Uexküll. According to von Uexküll, every animal exists within a "functional circuit" (*Funktionskreis*), or environment, that is constituted through the interaction of that species with the outer world (*Außenwelt*). This interaction is predetermined, he argued, through the specific organic structure of the animal, insofar as this structure only allows it to select certain stimuli (*Reize*) from the outer world and to respond to them in certain ways. However, these responses that the animal gives to the stimuli it receives have, again, specific effects on the outer world, so that, as a result, "a closed circuit" (von Uexküll 1973, 150) of stimuli and responses emerges. This was a proto-cybernetic idea, on which theories of ecology that emerged in the 1960s built. Plessner, however, was not especially interested in an ecological systems theory, but adhered to the older stage model of the organic world. Like Scheler, his main goal was to explain and justify the superior status of humans vis-à-vis the rest of the organic world. With reference to von Uexküll, he introduced the concept of positionality, arguing that organic beings actively "position" themselves in space and time. The human, in Plessner's view, is characterized by "excentric positionality", which means that, in contrast to plants and animals, humans are not bound to their environment but can reflect on it, and in this sense transcend the "here-now" (Plessner 2003, 364) in which they exist. Based on this assumption, Plessner claimed that the human being cannot rely on its natural instincts but is condemned to "natural artificiality", i.e. has to actively shape its environment and "live crabwise through artificial things" (384). Technology and the production of "artificial things" here emerge as the ultima ratio of the human.

A technological and productivist orientation of this kind also played a central role in Gehlen's theory of "anthropo-biology", which was published in 1940 under the title *Man: His Position and Nature in the World*. Gehlen claimed he could explain the exceptional status of the human solely by reference to biological knowledge, and thus avoid the dualistic structure that still prevailed in the accounts of Scheler and Plessner despite their attempts to overcome it. He drew in particular on the theory of "neoteny", formulated by the anatomist Louis Bolk, according to which the adult human preserves some foetal states. Gehlen conceived of the human in contrast to other animals as not being fully developed but rather a "deficient being" whose organs are largely unspecified and open to being used for various purposes. Like Plessner, Gehlen argued that the human is not bound to a certain environment and not determined by instincts but rather has to actively shape its living conditions. In addition, he considered the human to be shaped by a "constitutional excess of drives", i.e. an excessive energy that needs to be regulated and transformed. "Once the formation has succeeded," Gehlen claimed, "the constitutional excess of drives delivers an almost inexhaustible vectored energy which is transformed into activity, into labour", and which is geared towards achievements and products that "have the completely super-human capacity to dissolve as material into higher products" (Gehlen 1993, 428). Gehlen took the technological-productivist imperative even further than Plessner. He not only argued that the human has to build a specific mode of self-regulation and self-domination through which the ability to work and transform the nonhuman world is intensified, but also articulated subjectivation in terms of *Zucht* – which, in German, can mean both breeding and discipline – and *Führung* (leadership, command). Gehlen's support for the transformation of the subject's "excessive" energy certainly has sexual connotations, and his call for a regulation of sexual desire according to a certain "breeding ideal" alludes to Nazi eugenics.[9] Keeping "sexual fantasy (*Geschlechtsphantasie*)", as he put it, "faultless and fastidious" was, in Gehlen's view, an important task because he assumed that "the next generation" directly depended on "the quality of the image of the species" (*Gattungsbild*) (439) or the "breeding image" (*Zuchtbild*) (733) that shapes this fantasy. As this extreme example makes clear, notions of human exceptionalism reproduced the hierarchical views that structured the understanding of the organic world with respect to the subject itself, i.e. the regulation of its emotions, desires and needs. Just as Horkheimer and Adorno's analysis of the mythological figure of Odysseus revealed, the domination of "inner" and "outer" nature are inextricably linked to each other and to the whole web of societal relations of domination.

In clear contrast to the early nineteenth century, race and sex did not figure as central epistemic objects in the philosophical anthropologies formulated in the 1920s to 1940s. However, a close reading of the texts of Scheler, Plessner and Gehlen reveals that their accounts, too, were racialized and that they conceived of the human as masculine. Scheler, for example, claimed that the "plant principle" prevails in women and the "whole of (non-Jewish) Asia" (Scheler 1976, 15), suggesting that people from this part of the world, like women, cannot make it to the highest stage of full human development. Plessner in addition banished women from the sphere of artificiality – the sphere of true humanness – by arguing that they are particularly close to nature, while Gehlen claimed that the essential objects of human drives are "possessions, food, women" (cited in Klinger 1989, 195). So even if the anthropological narratives of Scheler, Plessner and Gehlen clearly differ from the theories of the late eighteenth and early nineteenth centuries – in particular with respect to their productivist overtones – their accounts of human exceptionalism were still gendered, racialized and based on hierarchical understandings of organic kinship.

Beyond the "Empirico-Transcendental Doublet"? Earth, Life and the (Non)Human

As we know, the figure of the human has not vanished, as Foucault predicted, although it has undergone significant change. It seems as if the nonhuman turn has initiated a shift from what Foucault called the "empirico-transcendental doublet" into an "onto-ethical" figure. This means that, in contrast to the history of philosophical anthropology, the human is no longer understood as a biological being that is, at the same time, the transcendental condition of the possibility of knowledge, but rather in terms of ontological claims that immediately translate into ethico-political ones.

In line with post-Kantian perspectives that have shaped the nonhuman turn, Rosi Braidotti defends an ontological position that builds on Gilles Deleuze and Félix Guattari's reading of Spinoza. She advocates a "vitalist materialism" and conceives of "matter as vital and self-organizing" (Braidotti 2013, 56). This ontological claim is the source of her ethico-political perspective. *Zoé*, as Braidotti puts it, "stands for generative vitality" (60) and is "the transversal form that cuts across and reconnects previously segregated species, categories and domains" (ibid.). The ethico-political imperative that follows from this, according to Braidotti, requires us to affirm this force in the name of a "zoé-centric egalitarianism" (ibid.). The main ethical attitude that is required from the human subject is to surrender to *zoé* and develop a kind of *amor fati*, a love of fate, "as a way of accepting vital processes and the expressive intensity of a Life we share with multiple others, here and now" (190). Latour, in contrast, opts for a different onto-ethical attitude. Based on his diagnosis of a "profound mutation of our relation to the world" (Latour 2015, 8), he introduces a distinction between the Earthbound and Humans. "Whereas Humans are defined as those who take the Earth, the Earthbound are taken by it" (251). Here, Latour not only re-evokes the patriarchal notion of a female mythological power – Gaia, or the Earth[10] – to which individuals are and remain bound by claiming that "in both cases the Earth is still the Mother of their law" (ibid.), but also adopts Carl Schmitt's distinction between friend and foe, and despite his explicit departure from Schmitt's aggressive geopolitics, the ethical attitude that Latour favours is basically one of enmity. Donna Haraway has commented that "Schmitt's enemies do not allow the story to change in its marrow; the Earthbound need a more tentacular, less binary life story" (2016, 43), i.e. one that acknowledges multiple lines of conflict without advocating aggression. These amalgamations of ontological and politico-ethical claims that characterize both Latour's and Braidotti's positions not only re-evoke traditional patriarchal notions of masculine aggression and feminine meekness and equanimity; the ontological turn to which they contribute also downplays epistemological perspectives. Thus, instead of scrutinizing the historico-epistemological conditions under which the ontological and ethico-political claims are formulated, and in particular the intellectual history of the figure of the human to which the nonhuman turn refers (albeit in a critical mode), parts of this troubling history are reiterated.

This holds in particular for the unacknowledged status attributed to techno-scientific knowledge in Latour's and Braidotti's accounts.[11] For Latour's argument, the reference to James Lovelock's bio-cybernetic Gaia theory is central. In contrast to Galileo's helio-centrism, a new geo-centrism is needed in the light of climate change, Latour argues. Lovelock is a crucial ally in his endeavour because, "Lovelock, lowering his eyes from Mars in our direction, in effect diminished the similarity between all the other planets and this so peculiar Earth that is ours" (Latour 2015, 78). It follows, Latour concludes, that "the place of the action is here below and right now" (81). Braidotti, meanwhile, argues that "the classical emphasis on the unity of all matter, which is central to Spinoza, is reinforced by an updated scientific understanding of the self-organizing or 'smart' structure of living matter" (2013, 57). In contrast to Latour, she considers the "Gaia hypothesis" problematic because it is "based on a social constructivist dualist method" (84), which means "that it opposes the earth to industrialization, nature to culture, the environment to society, and comes down firmly on the side of the natural order" (84–85). Instead, Braidotti repeatedly claims that her understanding of vital matter is "supported by new advances in contemporary biosciences, neural and cognitive sciences and by the informatics sector" (57). Like the theoretical figurations of the human in the narratives of philosophical anthropology, these present articulations of the (non)human are thus shaped by specific, historically contingent scientific knowledges, while the epistemic status attributed to the sciences remains unaddressed.

With regard to the critique of extractivist nature relations and human exceptionalism, this is particularly troubling, since the notion that matter itself is active and productive – which is central to the new materialism and the nonhuman turn broadly conceived – coheres in a paradoxical way with bioeconomic projects that seek to capitalize on the generative and regenerative processes of living beings of all kinds. As Jesse Goldstein and Elizabeth Johnson have argued, recasting "nonhuman life [...] as an active and 'wise' producer" (2015, 62) resonates with bioeconomic projects that seek to make the generative and regenerative capacities of living beings technologically and economically accessible. The attempt "to unseat Baconian conceptions of nature as a passive dominated object, offering instead a reconceptualization of nature as a realm of active, collaborative participants" (74–75) is thus not necessarily a step towards more ecological nature relations. On the contrary, highlighting the "agency" of species of all kinds and their overall interrelatedness can also give rise to new forms of extractivism that are meant to sustain economic growth after the end of the age of fossil fuels. "New natures", Goldstein and Johnson conclude, "can be and are tortured in new ways" (61). The question of whether critiques of human exceptionalism that are formulated in the context of the nonhuman turn may align with a critique of extractivism or whether they tend to promote new bioeconomic, post-fossil forms of it thus seems to still be open for discussion.

A major difference between the theoretical figurations of the human that I have outlined in the previous parts of this chapter and Latour's and Braidotti's accounts lies in their explicit criticisms of sex and race. Braidotti's critique of anthropocentrism clearly follows a feminist trajectory. It builds on her understanding of *zoé* as a process of becoming that overturns dualistic and hierarchical forms of difference, first and foremost sexual difference. Sexism, racism and anthropocentrism, according to Braidotti, all result from the same logic, which subordinates the Other to the Self. "In the political economy of phallogocentrism and of anthropocentric humanism, which predicates the sovereignty of Sameness in a falsely universalistic mode," she argues, "my sex fell on the side of 'Otherness,' understood as a pejorative difference, as being less-worth-than" (Braidotti 2013, 80–81). Although this homology between different forms of power and domination neglects the question of how different forms of power and domination are connected, Braidotti's intersectional perspective clearly departs from the sexist and racist implications of the history of philosophical anthropology. The same holds for Latour, although he is obviously less explicit about gender and feminism. However, Latour, too, rejects the "Western Masculine Subject" (2015, 62) and its "courageous, violent, sometimes disproportionate dream of control" (ibid.), and envisions a transformation of the relation between the "former colonizers" and "the formerly colonized" (Latour 2018, 14), which entails "a new universality" (ibid.).

However, these criticisms of the colonial, patriarchal and racializing aspects that have shaped the figure of the human since the late eighteenth century do not solve the problem of subverted universalism. Despite his reflections on a "new universality", Latour mainly defends a particularistic understanding of interrelatedness, replacing the (false) universalism of the notion of the human with the idea of different "peoples". The human, "as a unified agent, as a simple virtual political entity, as a universal concept", he argues, has to be "decomposed into several distinct *peoples*, endowed with contradictory interests, competing territories, and brought together by the warring agents – not to say warring divinities" (Latour 2015, 122). Of course, Latour's notion of peoples does not map onto existing nations, and in his view the combatants in the "war of all against each […] may not just be the wolf and the lamb but also tuna and CO2, plant nodules or algae, in addition to the numerous human factions that disagree about almost everything" (227). However, Latour claims that the Earthbound, instead of directing their view to the infinite universe in which all differences and inequalities vanish, resituate themselves on earth and within their respective relations of kinship. He conceives of a new "system of engendering" in which "all agents, all the animated things raise questions about descendants are forebears", centred on "the question of how to recognize and insert oneself into lineages that will manage to last" (Latour 2018, 59). To "insert oneself into lineages", however, may result in new and not so new forms of power and domination as well as aggressive particularism, given that human "kinship" relations have for most of human history been shaped by patriarchal and ethnocentric structures, through which wealth and privilege have been transmitted in highly exclusionary ways.

Rosi Braidotti, in contrast, remains faithful to an egalitarian-universalist perspective, and envisions a new "'we' that is evoked and actualized by the postanthropocentric creation of a new pan-humanity" (Braidotti 2017, 39). This vision is certainly more inclusive than any concept of the human has ever been but repeats the problem of false universalism insofar as it just ignores the multiple differences and relations of power and domination that disrupt the continuum of humanity and of life on earth. The "we", in Braidotti's view, "actualizes a community" whose members are bound together "by the compassionate acknowledgement of their interdependence with multiple others, most of which [...] are quite simply not anthropomorphic" (ibid.). Both Braidotti and Latour thus neglect structures of power and domination within the human world that shape the modes of extractivism by introducing notions of organic kinship and earthly belonging through which they seek to overcome the "humanist" tradition, i.e. the narratives of philosophical anthropology. Their conceptual alternatives to universalism collapse into calls to merge into larger entities such as "lineages" or, in the case of Braidotti, a "we" that encompasses all living beings.

Conclusion

"Think we must, we must think," Haraway repeatedly claims, citing Isabelle Stengers and Vinciane Despret (Haraway 2016, 30). Concepts and theories orient individual and collective practices so that, if the history of extractivist gender and nature relations is to acquire a different trajectory, a transformation of concepts, narratives and theories is also needed. With regard to the concept of humanity, Haraway has made some suggestions for reformulations in her earlier article *Ecce Homo, Ain't (Ar'n't) I a Woman and Inappropriate/d Others* (2004). In this article, she introduces the notion of "non-generic humanity" (48) through which humanity is recast as "a figure outside the narratives of humanism" (49), and without doubt also outside those of philosophical anthropology. With reference to Gayatri Spivak, Haraway argues that "humanity is that kind of thing [...] that which we cannot not want" (ibid.). She refrains from any attempt to define the human, but envisions a mode of collectivity and relatedness that does not yet exist but can and should emerge through what she calls a "self-critical practice of 'difference,' of the I and we that is/are never identical to itself, and so has hope of connection to others" (48). From this point of view, "humanity" is neither a universalistic norm to which practices have to be adjusted nor a mere biological, empirical facticity, but rather indicates the possibility that modes of being and living together on a planetary scale might result from multilayered struggles, deliberations and transformations. “Humanity” here implies neither a homogenization of perspectives and experiences nor a socio-centric view, but rather stands for envisioning modes of horizontal association that do not erase difference and conflict. Shaping such modes of association does not mean that we should “insert” ourselves into new and not so new “lineages” (Latour) nor that we should surrender ourselves to the “generative vitality” of living matter (Braidotti) but rather that we should actively shape and reshape gender and nature relations. If we cannot succeed in overthrowing the humanist and anthropocentric vocabulary – as the very notions of the post- and nonhuman that re-evoke the human, albeit with a new prefix, suggest – this might have less to do with intellectual incapacities and more with a situation in which the problems that are articulated through these concepts have not yet been solved. At the end of his subtle discussion of Marx’s *Theses on Feuerbach*, Étienne Balibar concludes that Marx does not succeed in fully putting aside the notions of humanity and humanness, because such notions constantly emerge from the conflictual structure of the world. “Metaphysical notions of human essence”, Balibar argues, “are not only inherited from an ideological past, but also permanently reconstituted through processes that ‘sublimate’ social contradictions of all kinds” (2012, no pagination). Against this backdrop, the present re-emergence of the figure of the human, both in the narrative of the Anthropocene and in the context of the nonhuman turn, can also be understood as a theoretical sublimation, or as a symptom of unsolved problems and conflicts. The need and unfulfilled desire to transform current societal nature relations that are structured by gender and class relations, global inequalities and imperial geopolitics, as well as various forms of racism, and to reshape the relationships of humans with the nonhuman world thus seem to fuel the permanent proliferation of new conceptualizations of (post-)humanness. For critical ecofeminism and the environmental humanities, the deconstruction and critique of old and new constructions of “the human” in order to render sublimated conflicts, problems and possible perspectives intelligible therefore remains a timely project.

Notes

1 On these concepts and the associated controversies, see Haraway (2016), Moore (2016), Di Chiro (2017), Murphy and Schoering (2020).

2 Such a “great family of living things”, as Kant put it, could be conceived as originating from “a universal mother [*gemeinschaftliche Urmutter*]” or from the “womb of mother earth as it first emerged [...] from its chaotic state” (Kant 2008, §80, 248).

3 For a more detailed discussion of Kant’s concept of race, see Bernasconi (2002). Anderson and Perrin (2018) rightly stress the significance of race theories for the anthropological accounts of the nineteenth century, but solely focus on comparative anatomy and phrenology. However, as Kant’s infamous definitions of the concept of race make clear, the concept was also closely linked to notions of reproduction and heredity, and thus to racialized and gendered politics of reproduction (Lettow 2014).

4 For a more detailed discussion of Kielmeyer's understanding of the organic world and the place of "sex", see Lettow (2021b).
5 See Natania Meeker's chapter in this volume.
6 The earth sciences were closely intertwined with the development of biology and the discourse of anthropology. In the eighteenth century, the "emerging sense that the earth and its inhabitants have an extended history" (Bowler 2015, 96) was closely linked to various theories of generation and reproduction, and thus to attempts to understand the specificity of organic nature. For a more detailed discussion of the entanglements of geology, geography and the modern race discourse of the nineteenth century, see Lettow (2021a).
7 For instance, Steffens claimed that in the north, the human "shrinks together with the vegetation, a sad deformity. All mental capacity dies away like the living activity of nature, and inwardly paralyzed, like the world that surrounds him, he is enslaved by base needs" (Steffens 1822, 365). In contrast, "the heat of the tropical countries produces another deformity" that expresses itself as "fervid desire" or in the form of "dawning dreams of an easygoing life" (366).
8 "Emerging as a distinctive school of thought in the 1920s, philosophical anthropology took its point of departure from Jakob von Uexküll's revolutionary conception of 'environment' (*Umwelt*) as relative to the perceptual and behavioral capacities of each species" (Horn and Bergthaller 2020, 78). In Germanophone scholarship, philosophical anthropology has seen a revival in the context of debates on the cultural implications of biotechnologies during the 2000s. Recent discussion has connected it to posthumanism and the Anthropocene discourse (Bajohr 2020).
9 Gehlen's book was published in 1940 when he became a professor in Vienna and was charged by the German Ministry of Education with "reorganizing" the philosophy faculties in occupied Austria. In editions published after 1945, Gehlen eliminated the most direct references to Nazism, although he did still, for example, claim that his "general anthropology" is a propaedeutic to the study of race (*Rassenlehre*) (Gehlen 1993, 488). On the relation between Gehlen's involvement in National Socialism and its theoretical impact, see Klinger (2000).
10 In the Anthropocene discourse, the earth and the figure of Gaia are constantly evoked, although in different ways. Miriam Tola (2016) shows how Isabelle Stengers' notion of Gaia – in contrast to Latour's – converges with ecofeminist and feminist-materialist approaches.
11 Whereas the earlier anthropological theories referred to scientific knowledge as an "empirical" ground on which they constructed ideas of human exceptionalism, the relation to the sciences is much more complicated in the nonhuman accounts. For decades, science and technology studies (in particular Latour's actor–network theory) have successfully questioned the sharp distinctions between science and culture and between the sciences and the humanities. As a result, the distinction between the empirical and the transcendental has become porous, too, and it does not seem productive to re-establish these distinctions. However, this should not prevent us from scrutinizing the way in which present reconceptualizations of the human in the context of the nonhuman turn refer to scientific knowledge.

References

Anderson, Kay, and Colin Perrin. 2018. "'Removed from Nature': The Modern Idea of Human Exceptionality." *Environmental Humanities* 10 (2): 447–472.

Bajohr, Hannes, ed. 2020. *Der Anthropos im Anthropozän: Die Wiederkehr des Menschen im Moment seiner vermeintlich endgültigen Verabschiedung*. Berlin: De Gruyter.

Balibar, Étienne. 2012. "From Philosophical Anthropology to Social Ontology and Back: What to Do with Marx's Sixth Thesis on Feuerbach?" *Postmodern Culture* 22 (3). http://www.pomoculture.org/2015/06/10/from-philosophical-anthropology-to-social-ontology-and-back-what-to-do-with-marxs-sixth-thesis-on-feuerbach-2/.

Bernasconi, Robert. 2002. "Kant as an Unfamiliar Source of Racism." In *Philosophers on Race: Critical Essays*, edited by Julie K. Ward and Tommy L. Lott, 145–166. Oxford: Blackwell.

Bowler, Peter. 2015. "Theories of Generation and the History of Life." In *The Secrets of Generation: Reproduction in the Long Eighteenth Century*, edited by Ray Stephanson and Darren N. Wagner, 79–99. Toronto: Toronto University Press.

Braidotti, Rosi. 2013. *The Posthuman*. Cambridge and Malden: Polity Press.

Di Chiro, Giovanna. 2017. "Welcome to the White (M)Anthropocene? A Feminist-Environmentalist Critique." In *Routledge Handbook of Gender and Environment*, edited by Sherilyn MacGregor, 487–505. Oxford and New York: Routledge.

Foucault, Michel. 1994. *The Order of Things: An Archaeology of the Human Sciences*. New York: Vintage Books.

Gehlen, Arnold. (1940) 1993. *Der Mensch: Seine Natur und seine Stellung in der Welt*. Volume I. Frankfurt: Klostermann.

Goldstein, Jesse and Elizabeth Johnson. 2015. "Biomimicry: New Natures, New Enclosures." *Theory, Culture and Society* 32 (1): 61–81.

Haraway, Donna J. (1992) 2004. "Ecce Homo, Ain't (Ar'n't) I a Woman and Inappropriate/d Others: The Human in a Posthumanist Landscape." In *The Haraway Reader*, edited by Donna J. Haraway, 47–62. London and New York: Routledge.

Haraway, Donna J. 2016. "Tentacular Thinking: Anthropocene, Capitalocene, Chthulucene." In *Staying with the Trouble: Making Kin in the Anthropocene*, 30–57. Durham, NC: Duke University Press.

Hausen, Karin. 1976. "Die Polarisierung der 'Geschlechtscharaktere' – eine Spiegelung der Dissoziation von Erwerbs- und Familienleben." In *Sozialgeschichte der Familie in der Neuzeit Europas*, edited by Werner Conze, 363–393. Stuttgart: Klett.

Honegger, Claudia. 1991. *Die Ordnung der Geschlechter: Die Wissenschaften von Menschen und das Weib*. Frankfurt: Campus.

Horn, Eva and Hannes Bergthaller. 2020. *The Anthropocene: Key Issues for the Humanities*. London and New York: Routledge.

Kant, Immanuel. (1775) 2007. "Of the Different Races of Human Beings." In *Anthropology, History, and Education,* edited by Günter Zöller and Robert E. Louden, 82–97. Cambridge: Cambridge University Press.

Kant, Immanuel. (1790) 2008. *Critique of Judgement*, translated by J. C. Meredith. Oxford: Oxford University Press.

Kielmeyer, Carl Friedrich. (1793–94) 1938. "Ideen zu einer allgemeinen Geschichte und Theorie der Entwicklungserscheinungen der Organisationen." In *Natur und Kraft: Kielmeyers gesammelte Schriften*, edited by Fritz-Heinz Holler, 102–194. Berlin: W. Keiper.

Kielmeyer, Carl Friedrich. (1793) 2021a. "On the Relations between Organic Forces." In *Kielmeyer and the Organic World: Texts and Interpretations*, edited by Lydia Azadpour and Daniel Whistler, 29–49. London: Bloomsbury.

Kielmeyer, Carl Friedrich. 2021b. "Ideas for a Developmental History of the Earth and Its Organisations: Letter to Windischmann, 1804." Translated by Iain Hamilton Grant. In *Kielmeyer and the Organic World: Texts and Interpretations*, edited by Lydia Azadpour and Daniel Whistler, 65–67. London: Bloomsbury.

Klinger, Gerwin. 1989. "Freiheit als 'freiwillige Aufgabe der Freiheit': Arnold Gehlens Umbau des Deutschen Idealismus." In *Deutsche Philosophen 1933*, edited by Wolfgang Fritz Haug, 188–218. Berlin: Argument.

Latour, Bruno. 2015. *Facing Gaia: Eight Lectures on the New Climatic Regime.* London: Polity Press.

Latour, Bruno. 2018. *Down to Earth: Politics in the New Climatic Regime.* Cambridge and Malden: Polity Press.

Lettow, Susanne, ed. 2014. *Reproduction, Race, and Gender in Philosophy and the Early Life Sciences.* Albany: SUNY Press.

Lettow, Susanne. 2021a. "Politicizing the Geological: Earth and History in Modern Philosophical Race Discourse." *Critical Philosophy of Race* 9 (1): 27–47.

Lettow, Susanne. 2021b. "Reproduction, Production and the Earth: The Place of Sex in Kielmeyer's 'Economy of the Organic World'." In *Kielmeyer and the Organic World: Texts and Interpretations*, edited by Lydia Azadpour and Daniel Whistler, 171–186. London: Bloomsbury.

Moore, Jason, ed. 2016. *Anthropocene or Capitalocene? Nature, History and the Crisis of Capitalism.* Oakland: PM Press.

Murphy, Michael Warren and Caitlin Schroering. 2020. "Refiguring the Plantationocene: Racial Capitalism, World-Systems Analysis, and Global Socioecological Transformation." *Journal of World-Systems Research* 26 (2): 400–415.

Oken, Lorenz. (1805) 2007. *Die Zeugung.* Weimar: Hermann Böhlaus Nachfahren.

Plessner, Helmuth. (1928) 2003. *Die Stufen des Organischen und der Mensch: Einleitung in die philosophische Anthropologie.* Frankfurt: Suhrkamp.

Scheler, Max. (1928) 1976. *Die Stellung des Menschen im Kosmos.* Bern and Munich: Francke Verlag.

Schiebinger, Londa. 1991. *The Mind Has No Sex? Women in the Origins of Modern Science.* Cambridge, MA: Harvard University Press.

Steffens, Henrik. 1822. *Anthropologie.* Vol. 2. https://reader.digitale-sammlungen.de/de/fs1/object/display/bsb10255567_00005.html.

Tola, Miriam. 2016. "Composing with Gaia: Isabelle Stengers and the Feminist Politics of the Earth." *PhaenEx* 11 (1): 1–21.

von Uexküll, Jakob. (1920) 1973. *Theoretische Biologie.* Frankfurt: Suhrkamp.

12 Speculative Ecologies

Salmon Farming and Marine Microplastics as Slow Disasters

Sven Bergmann

Ecological transformation and destruction happen within specific temporalities and spatialities. Studies on waste, pollution and toxicity are often confronted with uncertainty, speculative comprehension and unpredictable temporal effects (Auyero and Swistun 2008; Murphy 2006; Schrader 2010; Shapiro 2015). In order to address the specific temporality of ecological destruction and its consequences, various authors have suggested that these disasters and the violence they engender be described as "slow" (Knowles 2014; Nixon 2011). A characteristic of a "slow disaster" is that its effects often unfold much later, sometimes in other places or in other organisms – which gives it a speculative dimension. Knowledge production is therefore complicated, raising questions of responsibility and environmental justice in highly contested fields.

The two phenomena described in this chapter – the effects of unregulated salmon farming in southern Chile and the global accumulation of microplastics in the oceans – are analysed as slow disasters with speculative temporalities and a still unpredictable environmental impact.[1] Moreover, they are particularly difficult to study because of their effects in the fluid medium of the ocean. Within the marine environment, the floating, accumulation and sedimentation of anthropogenic substances (e.g. microplastics, chemicals, antibiotics) call into question the modern distinction between nature and culture. As a result of these processes, new entities and life forms emerge in the sea. Deeply gendered assumptions of the purity and pristineness of nature that sometimes underlie environmental approaches have, consequently, been criticized from the perspective of feminist technoscience (Liboiron 2016; Murphy 2008; Nash 2008). In view of the growing ecological challenges of our times, the question arises of how to develop a new politics and ethics of care (Martin, Myers and Viseu 2015; Puig de la Bellacasa 2011). Caring as a "speculative commitment" (ibid., 96) can make a difference in conceptualizing ecologies of gender and its entanglements with the more-than-human.

In 1978, London punk singer Poly Styrene wrote a song about a world that artificially shone in luminous colours: "I clambered over mounds and mounds/Of polystyrene foam/And fell into a swimming pool/Filled with fairy snow" (quoted from Brown 2011, 461). These lyrics from her band

DOI: 10.4324/9781003023319-17

X-Ray Spex's second single "The Day The World Turned Day-Glo" may have sounded dystopian at the time. However, several decades later I have walked along beaches in the south of Chile that were full of expanded polystyrene, flotsam from local shellfish farming. Meanwhile, a study from France has demonstrated that, in addition to microplastic pollution of bodies of water, there is also atmospheric fallout of microparticles, especially fibres from plastic textiles such as fleece sweaters, albeit perhaps not in such neon colours these days. For Poly Styrene, the obsessive preoccupation with plastics was not only a criticism of the capitalist mode of production, consumption and environmental destruction as the lyrics quoted above suggest, but also a rejection of everything and anything considered authentic (Brown 2011).[2] This tension between a critique of environmental pollution and a simultaneous uneasiness with environmentalist discourse such as notions of a pure or untouched nature is of central importance to this essay. It is at these interfaces that certain premises of environmental protection can be challenged by means of postcolonialist and queer feminist approaches, while still allowing us to discuss how we can resolve ecological problems such as global warming, ocean acidification and pollution.

In this chapter, I will discuss two different cases of environmental problems, both of which occur in marine and coastal regions: first, harmful algal blooms and second, the high concentration of microplastics in the ocean. Although the temporal effects of land-based forms of pollution are often difficult to tackle, this becomes an even more challenging task within marine and maritime environments. In the first case, I present the regional phenomenon of a recurrent toxic algal bloom in southern Chile, which is discussed in connection with large-scale aquaculture, mainly salmon farming. The second case is about a much more dispersed form of marine pollution, namely the ubiquitous global accumulation of plastics in the oceans, whose effects on ecosystems are more unpredictable. I will discuss both cases together as forms of speculative ecologies that challenge conventional notions of nature, culture and temporalities, and therefore also typical ways of negotiating or solving these problems. Before I present the two case studies in more detail, I will provide a short excursus on the theorization of indeterminable futures of environmental change.

Speculative Ecologies: Temporalities of Slow Violence

> Leaks, leachates, chronic exposures, decaying infrastructure, ocean acidification, and climate change are slow disasters. Their harms are too amorphous and do not register easily within regulatory metrics. Their scales are too large or too small to observe directly. They are not readily manifested.
>
> (Liboiron, Tironi and Calvillo 2018, 338)

Whereas concepts of environmental justice have mostly targeted spatial and distributional aspects of environmental damage (Sandlos and Keeling 2016, 8), slow violence brings larger historical processes of political ecology and future potentialities into the picture. Pollution and toxicity are often a form of "slow violence", a kind of violence that is difficult to recognize because it is often invisible and its effects delayed. The concept of slow violence was defined by literary scholar Rob Nixon as "a different kind of violence, a violence that is neither spectacular nor instantaneous, but rather incremental and accretive, its calamitous repercussions playing out across a range of temporal scales" (2011, 2). Although the agents of slow violence are sometimes portrayed in Nixon's writings as utterly inexorable, the concept's strength lies in its addressing environmental justice on a more temporal scale, thus encompassing historical, colonial and, in turn, future dimensions of temporality. The effects of slow violence are not merely historical, but extend to potential futures and future generations: "Slow violence presents us with a political geography of deferred environmental threats, where violence is out-sourced – not only to the Global South – but also to a Global Future" (Davies 2019, 2). Whereas environmentalism in the North is considered a more "White" and middle-class issue, and is often detached from other social struggles, or even regarded as part of a Western imperial project, ecological politics is much more intertwined with social conflict in other parts of the world. Thus, the pollution of land and water can be noted in the continuity of (settler) colonialism because it restricts people's access to these resources and endangers their lives. Consequently, environmental activism in other parts of the world is often a highly risky, even life-threatening endeavour. Moreover, anthropologist Joseph Masco writes that the gap in time between environmental events and the recognition of their long-term repercussions is "a major psychosocial achievement of the industrial age" (2015, 153), further reinforcing slow violence's capacity to affect both the past and the future, both unpredictably and over multiple generations. For this I introduce the term "speculative ecologies" in this chapter to refer to the vagueness of changes in ecosystems and their uncertain temporality, for example, when fishermen might have to reckon with a renewed dangerous algal bloom.[3] Particular ecological disasters such as residues from mining or toxic inputs into the ocean often have consequences years or decades later, and therefore risk affecting future generations. These phenomena require different strategies and forms of political representation. The vagueness of research results in this area also leads to a problem of knowledge transfer and, furthermore, presents a target for sceptics or deniers of ecological problems. Knowledge production in these areas is thus highly politicized. The relevance of ethnographic research lies, among other things, in its potential to demonstrate which local actors, objects and discourses are excluded in these negotiations, and what kind of indigenous knowledge is rendered invisible by powerful translation strategies. According to María Puig de la Bellacasa, "engaging with care requires a speculative commitment to neglected things" (2011, 85). Waste and residues are something which cannot be neglected; we must deal with them: "Instead of WM's (waste management) ultimate aim of making waste disappear, caring for waste necessarily recognizes that waste is not going to go away, no matter how much we wanted it or planned it" (Ureta 2016, 5). This problem is even more complicated within the marine environment: the pollution of the sea is difficult to analyse due to the mobility and dispersion of its inputs. This further complicates the production of knowledge and truth claims, rendering this a difficult undertaking for local actors, as will be shown in the empirical case studies below.

Outbreak of *Marea Roja* in a Salmon Farming Region in Southern Chile, 2016

Between February and April 2016, harmful algal blooms affected the coasts of Chiloé, an island in Chile's southern region with some 168,000 inhabitants, a third of whom belong to the Huilliche indigenous group. First, there were fish deaths in salmon farms due to a harmful algal bloom of *Pseudochattonella verruculosa*. Around 50,000 tonnes of salmon suffocated to death in their cages (Cabello and Godfrey 2016). Later, in April and May, another harmful algal bloom called the *marea roja* (red tide) extended from the Chiloé Interior Sea (where most aquaculture is located) towards the north and the open Pacific Ocean. The dinoflagellate *Alexandrium catenella*, which produces a paralytic neurotoxin, had already caused smaller red tides in Chile, but the outbreak in 2016 provoked a local state of emergency. A short time after the outbreak, locals observed dead animals ranging from sardines and mussels to penguins, and even whales washed ashore along the beaches. Prior to this, during March 2016, about 4,600 tonnes of decaying salmon had already been dumped into the open Pacific Ocean some 75 miles away from the coast of Chiloé (Cárdenas 2017).

In recent decades, Chile has become the second-largest exporter of salmon in the world. In Chiloé and other southern regions, salmon farming and related industries such as fish feed factories have become big business, mostly dominated by transnational companies. Dumping large quantities of infected and dead salmon into the sea is emblematic of how, even in the twenty-first century, the ocean is seen as a vast resource for capitalist extraction in Chile, or a huge dumping ground for waste residues like dead fish. Either way, it aroused great indignation among the Chilote people (as the inhabitants of Chiloé are known). As Gustavo,[4] an environmental activist whom I met in November 2016, half a year after the incident, said: "And then we had a crisis. An environmental crisis that became a political, economic and social crisis, then a global crisis for the whole archipelago".

Subsequently, in May 2016, one of the biggest social mobilizations in the history of Chiloé took place, lasting for 16 days. It was started by fishermen fighting for compensation, but was then followed by a huge wave of solidarity from Chiloé's wider population. While the fishermen were primarily demanding compensation, the strike brought also to the fore social movements angered by years of structural neglect of the region. Environmental problems were also discussed in relation to the exploitative salmon farming that had been practised in the region for years. The mass mobilizations finally stopped on 19 May when the fishermen agreed to compensation of 750,000 pesos (about 1,500 US dollars) per family. Not all of them agreed with the deal struck by the negotiators of the fishermen's organizations. As Irene, an anthropologist who works in public health, remarked:

> There are people who are working a bit on it, they are working after the events in May and the end of the mobilizations, who wanted to continue working on the issue of the social movement, I don't know if you've heard of it, called Chiloé está privao. [...] It clearly arose because of the May crisis and the fact that things were being resolved and everyone went home and people thought that something could still be done about it.

Chiloé está privao is a Chilote expression denoting fury: Chiloé is enraged. During those May days, many other issues came to prominence: environmental questions raised by the longstanding effects of salmon farming on the island, as well as questions about the lack of infrastructure on the island and the desire for more cultural autonomy.

The Chilean government has neglected the region for years. The question of education and universities is of crucial relevance for the Chilote activists because there is no scientific institution on the island to undertake research on the effects of harmful algal blooms and their relevance to salmon farming. The production of knowledge about the causes and effects of the 2016 algal bloom thus came to be contested between the Chilean government and the operators of salmon farms, on the one hand, and the new social movements in Chiloé, on the other. The protestors on the island had claimed that the industry's dumping of dead salmon into the ocean had caused the massive red tide. When I visited Chiloé in November 2016, half a year after the events in May, the government had already published a final report (*Informe final*) about the harmful algal bloom. According to the report, the toxic algal bloom was a natural event unrelated to salmon farming in the region, which had been subject to a "series of climatic and oceanographic anomalies during the summer and autumn of 2016 that explain the proliferation of phytoplankton, including harmful species" (Buschmann et al. 2016, 53). The final report also assured readers that salmon was dumped far away from land, and that the direction of drift was not towards the coast due to the wind. However, according to independent media sources, researchers and local fishermen, it seems rather doubtful that the outbreak and its development have nothing to do with the massive environmental transformations of the region brought about by the salmon industry, including the discharge of antibiotics and waste into the water, the huge quantity of dead salmon dumped in spring 2016 being one example. Other marine scientists believe that the red tide is a consequence of higher water temperatures due to the weather phenomenon of El Niño. Nevertheless, they also acknowledge the high nutrient loads (resulting from the enormous waste from salmon farming) as a likely contributing factor (Mardones, Müller and Hallegraeff 2017). Following the events in May, the production of knowledge about the causes and effects of the 2016 algal bloom thus came to be contested between the Chilean government and the new social movements in Chiloé. The final report on the *marea roja* was strongly criticized by activists on the island. Fabian, who was working at a municipal environment office in Chiloé, said:

> The issue of the red tide is complex because I think it is a combination of different factors. It has to do with climate change, but also with the input of nutrients by cultivation centres. The salmon farms are to be blamed, for example because they dump the dead salmon. However, it is strange that, on the one hand, the government says that "nothing happened" and, on the other, Greenpeace says the complete opposite. Who is right? [This is] because there is no good local university here, a scientific institute that is not corrupt and does not have ties to interested parties.

As can be seen from quotes like this, the environmentalists do not deny the complexity of the reasons for the outbreak of the *marea roja*, but they mistrust the commission appointed by the government because it does not attach any particular importance to temporality, the decades of environmental transformation and destruction caused by salmon farming. Fabian also complained of an infrastructural lack, with no local institutions and universities able to study the problem from the bottom up. Marcela, another member of an environmental organization, attempted to assess the current phenomenon in the context of the years of economic and ecological change in the region:

> This report is only concerned with the (possible) relationship between the dumping [of salmon] and the occurrence of the red tide. After our review of the literature, we concluded that the dumping of dead salmon did not have a major impact on the emergence of the red tide. In contrast, the report silences what has happened here in the last thirty years, the dumping of an incredible amount of nutrients. [...] To produce one tonne of salmon, 1.2 tonnes of fishmeal are necessary, of which a certain amount (75%) sediments on the seafloor, representing a tremendous input of nitrogen and phosphorus. An estimated annual amount of 667,000 tonnes of nitrogen and phosphorus – [imagine that] over thirty years.

All interviewees emphasized the large quantity of inputs that had accumulated in the sea, which could have a certain temporality, as Marcela extrapolates from the nitrogen and phosphorus inputs from salmon farms. It is therefore one thing for the government to claim that dumping salmon has no effect on algal blooms, but another, and far more serious, not even to acknowledge the effects of over 30 years of salmon farming in the region, as vividly described by Marcela.

The historian Andrew Gerhart has studied the occurrence of the Chilean salmon flu that caused such dramatic damage to the salmon industry and job losses for workers in 2007. According to Gerhart, "salmon farms are petri dishes, and the fjords of southern Chile, a laboratory" (2017, 738). He demonstrates this by comparing the use of antibiotics, which is some 300–400 times higher in Chile than in Norway, the country's main rival in salmon farming (727). Gerhart emphasizes the role of materiality and non-human agents as traces, drawing on anthropologist Gastón Gordillo's notion of ecological rubble to characterize the ecological debris in a marine space (729) that is supposed to be flushed out to sea. Nonetheless, the ecological rubble might have interacted with the El Niño phenomenon and the dinoflagellate *Alexandrium catenella* (738). Knowledge production is complicated and contested in environments of uncertainty.

Contested Knowledge Production in Environments of Slow Violence

As science and technology studies (STS) researcher Astrid Schrader has demonstrated in her study on the scientific controversy over *Pfiesteria piscicida*, studying harmful algal blooms is a difficult and complex endeavour. Even after two decades of research, this dinoflagellate could neither be confirmed nor ruled out as the causative agent for fish deaths on the US Atlantic coast. Drawing on Derrida's notion of spectre (as elaborated in *Spectres of Marx*), Schrader shifts attention from the uncertainties in human knowledge production to the notion of indeterminacy and what she calls "phantomatic ontologies", writing: "Phantoms do not emerge 'as such'; they appear as traces and are associated with specific matters of concern" (2010, 279). Thus, the dinoflagellates' intra-actions with their environment depend on context: if there are no fish, there is no fish killer. Furthermore, toxicity might not be independent of the environmental conditions in which it emerges. It is, rather, indeterminate whether the dinoflagellates' "toxicity is occurring for the first time or whether it is coming back" (293). Nevertheless, phantoms leave traces.

People in Chiloé live in conditions that the Argentinian ethnographers Débora Swistun and Javier Auyero term "toxic uncertainty", characterized by a "'labor of confusion' produced by socially consequential institutions" (2008). In the case of Chiloé, powerful actors such as the neoliberal Chilean state and multinational companies produce this labour of confusion; the scientists and universities that cooperated on the final report on the red tide were regarded by activists from Chiloé as being too involved with the Chilean state, and as also contributing to the confusion. Marcela emphasizes that there had been an input of substances that had accumulated for years, and so this input might have an impact on the ocean. Nevertheless, because of inter- and intra-actions with other complex marine phenomena, this raises uncertain and unpredictable temporalities with regard to when an algal bloom or comparable incident will happen again. People on the island know that there will be future outbreaks, but do not know when. They must deal with uncertainty and indeterminacy in their living conditions and in their perspectives on the future.

The example of living with indeterminacy in Chiloé and undecidable temporalities of potential outbreaks of salmon flus and red tides demonstrates how these uncertain temporalities together form a complex that I introduced earlier with Nixon's concept of "slow violence", which focusses on socio-environmental impact "across a range of temporal scales" (2011, 2) including "the temporal distance between short-lived actions and long-lived consequences" (41). What is striking in the final report is that although it does acknowledge the incident, it does not factor in the structural socio-ecological transformation of the region for decades, the devastating impacts this has had on the marine environment or the disastrous working conditions in salmon farming, which lead to several, often fatal diving accidents each year. The responses from national institutions "seem short-term oriented and uncertainty undermines their ability to address future environmental challenges" (Arriagada 2019). Although the red tide of 2016 may not have been primarily caused by salmon farming, this environmental disaster sheds light on the catastrophic conditions within the region's salmon farming industry, which might produce other delayed effects. Even without sudden disasters such as a dangerous algal bloom, the breeding of salmon in tiny cages together with their excrement and the high mortality rates and uncontrolled addition of antibiotics all remain an ongoing ecological challenge, representing a form of slow violence enacted upon the region (Iizuka and Katz 2011; Quiñones et al. 2019). However, in contrast to Nixon's largely opaque and invisible version of slow violence, there were many witnesses in Chiloé to the disaster, and a social movement emerged that triggered a genuine political crisis for the Chilean state. As Eric Thomas argues, fishermen and activists in Chiloé created a form of "catastrophization from below" throughout their social mobilization (2018, 208) that transformed a "regional *environmental* crisis into a national *humanitarian* crisis" (201) concerning the responsibility of the Chilean state.[5]

In order to discuss the situation in Chiloé, it is helpful to examine the slow violence exemplified by years of contamination of the coastal region by salmon farms in relation to structural neglect of the island and forms of epistemic violence. Thom Davies argues that there are many stories of slow violence, but they are generally discounted. He draws on Gayatri Spivak's concept of epistemic violence to describe the ignorance regarding "local claims of environmental injustice" (Davies 2019, 13), which creates a divergence between official and local accounts of socio-environmental problems. As a result of structural dominance, the official reports can usually rely on the cooperation of scientific institutions.[6] In the case of Chiloé, the social movement was successful in gaining compensation for the fishermen. However, as far as knowledge production is concerned, the movement is still struggling against epistemic violence and hegemony in order to combat the environmental damage – and slow violence – that salmon farming in the neoliberal and extractivist scheme has caused and will continue to cause, unless it is done in a different way. The new environmental and social movements in Chiloé have advanced notions of autonomy and independence for the island.[7] New social movements like *Chiloé está privao* and Defendamos Chiloé are highly active on social media, strengthening a Chilote national identity and heavily drawing on the Huilliche indigenous identity. Remarkably, a conflict over the exploitation of natural resources has become a battle for cultural hegemony and autonomy in Chiloé, reframing pollution and primitive accumulation as a current form of colonialism.[8]

When I walked along the beaches of Chiloé's interior coast and photographed the huge quantities of expanded polystyrene – known locally as "plumavit" – and other plastic litter, I also found it interesting that plastics have not yet been considered in southern Chile as a potential vector in the spreading of diseases and algal blooms (in addition to dead salmon, uneaten fish pellets, antibiotics and fish faeces). Recent studies have already shown that plastic particles in the water can serve as a carrier of species capable of producing toxic algal blooms such as dinoflagellates (Casabianca et al. 2019). I will return to this idea later when I discuss bacterial habitats on plastics (the *plastisphere*). This now brings me to my second case study: the problematization of (micro)plastics in the sea.

The Ubiquity of Microplastics in the Ocean and Their Accumulation in Rapa Nui

Plastics in the ocean have become one of the most recognized environmental marine problems in the twenty-first century. In this case, scalability plays a considerable role. Media attention is drawn to the effects of larger and coherent plastic objects in the water, and so more to the immediately evident forms of harm or violence caused by this material in the water. The impact of macroplastics upon sea life is typically addressed by means of dramatic photographs of sea animals like seals or turtles entangled in fishing nets or six-pack rings, or of an albatross with a mass of plastics in its stomach.[9] By contrast, microplastics, which are invisible on the moving ocean surface, are incompatible by these representations. In the marine sciences, the term microplastics emerged around 2004 as a subclassification of plastics; a distinction was drawn between macroplastics (> 5 mm) and microplastics (< 5 mm).

Sometimes, microplastic particles are released into the sea in microplastic dimensions. One example is plastic pellets, which are the raw materials for plastics production. Most microplastic particles in the sea, however, originated from larger plastic parts that fragmented in the sea due to various factors, such as UV radiation. Marcus Eriksen et al. (2014, 7) emphasize that over 92 per cent of plastics in the ocean are less than 5 mm in size. Unlike macroplastics, the notion of microplastics points to relatively unknown effects that are less well studied, especially as regards the "long-term effects of continuous exposure" (Lambert and Wagner 2018, 1).

Moreover, recent research and monitoring has revealed a significant presence of plastics not only in the sea, but also in freshwater systems, in the soil and in the air. For example, a research team recently detected an atmospheric fallout of textile fibres in Paris (Dris et al. 2016). Even in its macroscopic form, plastic is less than easy to control and contain; in the form of microplastic, it seems all but ubiquitous. Tiny synthetic polymers and their accompanying chemicals can be found in the most remote regions of the ocean as well as in the soil, in the air, in plants and animals, and even in human bodies. Micro- and nanoplastics show the importance of scale when it comes to materiality and toxic entanglements.

Plastics have become ubiquitous in the sea. Yet there are some zones in the ocean where more plastics accumulate than in others. Hence, global plastic production and consumption and the material's unplanned afterlives can affect marine environments even in the most remote regions like the Arctic and Polynesian islands such as Hawaii and Rapa Nui.

On the Polynesian island of Rapa Nui (Easter Island), the global plastics crisis can be seen as if through a prism. Since the 1990s, the island has flourished as a tourist destination, and the litter accumulated as a result of tourism, including tonnes and tonnes of PET bottles, stretches the infrastructure of the island (population: around 7,000) to the absolute limit. Furthermore, the island is located near accumulation zones (plastic gyres) in the ocean, so that small to tiny plastic fragments are constantly being washed up on its shores. The island (annexed by Chile in the nineteenth century), located thousands of kilometres from the mainland and other islands, is thus experiencing the effects of the plastics crisis twice over. Sofía, an activist I met during a seminar about ocean conservation on the island in collaboration with Chilean scientists from ESMOI (Ecology and Sustainable Management of Oceanic Islands) and *Científicos de la Basura*, told me how she feels about the situation on the island:

> During the lectures of the last few days, several Rapa Nui said, "If we don't produce the rubbish why do we have to endure it?" Well, for our people, most of us have felt violated in different ways, by the state, by foreigners, by the majority, and we have no recourse, we can't do anything to stop it.

What Sofía is referring to here is a specific version of environmental injustice: a portion of the waste produced by human civilization in other parts of the planet washes up on the island's shores, and the local people bear the consequences far away from where the materials first entered the ocean. Moreover, this relationship intersects with colonial violence. Sofía mentioned the numerous *moai* that were transported to other parts of the world:

> Plundered, for here things must be clear. Some say that the one in the British Museum was exchanged for a blanket. Can you believe that my people at that time spoke English? Did they know what a blanket was? If I told you what a *moai* means, it's someone in my family, it's an ancestor of mine, am I going to exchange that with you? So that's why today [...] diversity must be respected, we must be respected and understood.

In the case of Rapa Nui, the fight for restitution of ancient heritage such as the *moai* can proceed hand in hand with struggles on environmental justice issues such as plastic pollution, as demonstrated by Ida Luz Hucke Atán (1958–2018), known as Mama Piru, who was a pioneer on the island when it came to recycling, organizing beach cleaning and raising awareness of environmental issues and decolonization. Polynesian cosmology, based on the idea of living with and from the ocean (Hau'ofa 2008), gives an expression to these practices that goes beyond environmentalism, engaging in what Marisol de la Cadena (2015) describes, with reference to Isabelle Stengers, as "cosmo-politics". In Rapa Nui, environmental protection and postcolonialism are thus in negotiation. Both practices are linked by a concern with preserving the island's environment, caring for its past and future. In the next section, I will discuss the relational and temporal aspects of plastic, a substance that is never alone, in the marine environment.

A Relational and Temporal Perspective on Chemical Regimes of Living

In the emergent scientific literature on plastics, as well as within environmentalism, the focus has often been solely on the plastics themselves, although they rarely travel alone. Rather, they bring with them a raft of chemical and biological relationships and entanglements. Plastics usually come with all manners of additives, such as flame retardants, plasticizers or even bisphenol A, a substance known to have hormone-modifying effects. In addition, because of plastics' hydrophobic surface, persistent organic pollutants (POPs) such as PCBs and DDT attach to waterborne plastics. These plastic–chemical relations introduce another perspective on plastics in the environment, namely, as an ecotoxicological problem.

Michelle Murphy coined the term "chemical regimes of living" to describe "the accumulated result of some two hundred years of industrialized production, such as coal-based energy of the nineteenth century, or petroleum and plastic processing of the twentieth century" (2008, 697). She identifies another characteristic of a chemical regime of living in the problem of "commodity violence". Murphy notes that industrial pollution and toxic emissions might be "detectable but nonetheless irrelevant to corporate accountability" (698). As seen in the case of Rapa Nui, the temporality of commodity violence is characteristically slow, with delayed effects. Following the arrival of plastics some 70 years ago as a successful post-Second World War consumer good, there is now a large remnant of these materials on the planet, which will remain here for an indefinite period of time. Plastics accumulate in the ocean and are washed up on coasts far away from the places where they are produced, consumed and put into the water, and hence far away from the responsibility of the plastic industry.

Nevertheless, the issue of plastics in the ocean has become one of the most popular environmental topics over the last years and has pushed other environmental issues into the background – until the climate movement gained momentum again due to the school strikes that began in 2018. However, many solutions proposed to the problem of plastics in the ocean are still very much driven by a land-based, macro-scale perspective. In the case of marine plastics, media representations evoke huge "plastic islands" that no research cruise has yet been able to confirm because microplastics are relatively invisible on the moving ocean surface (De Wolff 2014). Moreover, the proposed solutions tend to treat the ocean with land-based solutions for waste management, which neglect marine-specific scalabilities, and often do not reflect that, for instance, not every single-use plastic was created for leisure purposes. Thus, a ban on plastic straws might be barely an inconvenience for the majority of people but a real problem for disabled people who need the straws to drink. Here, perspectives from feminist STS and disability/crip studies are vital for an analysis of intersections and interdependencies, with a focus on power relations in technology and infrastructures (Wong 2019; see also Star 1990). According to Puig de la Bellacasa (2011, 100) these approaches have a "commitment to neglected things" – for instance, they pay attention to emerging standards such as the straw ban and focus on neglected and non-conventional perspectives.

Moreover, it is difficult to represent slow violence because of the specific temporalities and spatialities of ecological destruction. Plastics have slowly been accumulating on the earth and in the ocean in a highly dispersed manner. As Heather Davis (2015a, 234) writes, "the relationship between cause and effect often appears much later, or [...] in completely different organisms". For example, chemical components of plastics such as bisphenol A (BPA), which is classed as an endocrine disruptor, are observed at high rates in certain fish or seabird populations, but not in other species. Studies of contamination and long-term environmental and health impacts of ecological disasters must also contend with highly unpredictable effects.

Furthermore, it is not just new chemical relations with plastics that occur in the marine environment; there are also new biological entanglements with plastics, such as bacterial habitats on plastic particles that differ from other bacterial biofilms in the ocean.

The emergence of aggregations and interactions with microplastics is one of the most thought-provoking fields of study in this area. Whereas plastic tends to be represented as an alien intruder within environments, this neglects the ambiguous roles it also plays. Plastics in aqueous environments are a cause of disturbance, but are also a new habitat, a surface upon which new life forms are created. Plastics in water are never alone, but are colonized by diatoms and other microbes within a very short space of time. A team of marine scientists has termed this world of new life forms the plastisphere (Zettler, Mincer and Amaral-Zettler 2013). Although in these cases bacteria are doing what they have always done, namely, searching for durable habitats in a competitive environment so as to create biofilms, the role that an anthropogenic material might play in these processes is of considerable interest. The bacterial biofilm on plastic particles differs from bacterial habitats in the surrounding ocean. Hence, it can have a major impact on marine ecosystems, as the encroachment of the vibrio bacteria in the Baltic Sea demonstrates (Kesy et al. 2016). As plastics become increasingly entangled and aggregated with marine species such as plankton (Long et al. 2017), working with notions of purification or separation seems all but meaningless. I initially coined the term speculative ecologies for a problematization of these phenomena.

The long-term effects, potentials and/or risks on and for marine ecosystems created by these interrelations of synthetic and biological in the sea are difficult to assess. In an interview that I conducted with plastisphere researchers in 2016, Erik Zettler commented on possible effects that bacterial colonization of plastics used in aquaculture could have, since the plastisphere is also a habitat for pathogens such as several vibrio bacteria. In a more recent article, he and his colleagues see this as a potential avenue for research (Amaral-Zettler, Zettler and Mincer 2020); this, in turn, links together the two cases discussed above: harmful algal blooms and microplastics. Although plastics are frequently used in aquaculture facilities, the dispersion of pathogens and also microbes potentially responsible for harmful algal blooms (ibid.) has not yet been studied in depth. This leads me back to my own thoughts about the use of plastics in aquaculture in the penultimate section. To summarize, the plastisphere researchers say that:

> It is clear that taking hundreds of millions of metric tonnes of hydrocarbons out of the Earth's interior and producing refractory materials that are allowed to escape into aquifers and marine systems has set up an "experiment" for which we are only beginning to interpret the results. (ibid.)

Both the cases discussed in this chapter can thus also be seen as an ongoing "experiment with unpredictable effects" (Bergmann 2021).

Queering the Plastisphere?

Nature and culture cannot be easily separated, and we are still haunted by the "ghosts of natureculture", as STS scholar Banu Subramaniam puts it, "as notions of purity and authenticity haunt our imaginations of science and the natural" (Subramaniam 2014, 117), instead of our reflecting on our own involvement in these processes (141). For this reason, the aggregation of microplastics in the ocean and the emergence of new hybrid life forms on synthetic surfaces can serve as a means for reflection on human entanglements with plastic and the embeddedness of synthetic polymers within human environments, infrastructures and even bodies, all of which challenges nature/culture and human/non-human dichotomies.[10] Concepts from queer studies are helpful for questioning binary and anthropocentric logics of representation. Although these logics help us to understand and represent complexity and entanglements in emerging phenomena, they concurrently tend to keep the subject/object of analysis at a distance.[11] Hybrid environmental objects like the plastisphere have become "good to think with" within the social and cultural sciences, yet preserve a gap between fascination and on-the-ground engagement.

On account of its evasive and erratic character, plastic in the ocean calls into question notions of nature as pristine or authentic.[12] This renders it an interesting object of study because it confronts environmentalist thought with postcolonial concepts (see Nixon 2011, 233–262). Environmental discourse, such as that in Rapa Nui as described above, intersects with anticolonial and postcolonial practices, so that pollution can also be described as a form of colonialism (Liboiron 2021). In addition, it opens the floor for a discussion of queer temporalities, as Heather Davis has argued in an article on the queerness of the plastisphere. Davis introduces plastic as a kind of chemical kin: "a non-filial human progeny, a bastard child that will most certainly outlive us" (Davis 2015b, 232). This is reminiscent of Jack Halberstam's definition of queerness as a "potentiality of a life unscripted by the conventions of family, inheritance, and child rearing" (Halberstam 2005, 2). Davis aligns the potential of the plastisphere to Lee Edelman's notion of queer non-reproductivity. In the ocean, plastics and their related chemicals such as bisphenol A, are contributing to this kind of "technobacterial future" (Davis 2015b, 242). From a different perspective, Davis is searching for a "crip theory for the current biosphere" (243), drawing on Mel Chen's concept of queer toxicity. Chen's concept might here prove helpful in challenging traditional environmental rhetoric such as the doomsday clock. It has been five minutes before 12 my entire life; by way of contrast, toxicity can alternatively be framed as being "already 'here'" (Chen 2012, 218), as we are already chemically altered and living in times in which we must reconfigure and reflect upon concepts such as ruin, contamination and disturbance (Tsing 2015). Plastic as a symbol of the non-authentic, as used in the punk lyrics of X-Ray Spex, challenges notions of pure identity, narratives of pristine nature and concepts of a heterosexual and non-intoxicated temporality.[13] Concepts from queer theory help us to understand the specific temporality of plastics in a different way. Challenging the idea of nature as a pure category is not just a theoretical game; rather, it is about creating other focal points of concern and care: a "transnatural ethics" (Krupar 2012, 311) as a reorientation of environmental ethics that reinforce nature/culture binarism and that might reflect the "porosity and permutability with waste and the human" (*ibid.*), since waste such as plastic is an unmanageable material and does not simply disappear. With the introduction of plastics into the ocean and the amalgamation of microplastics with marine ecosystems (such as the plastisphere), new life forms emerge that transcend linear temporalities and create rather speculative ecologies.

Furthermore, forms of slow violence that occur in the ocean and primarily affect non-human species at present, and only temporarily impact upon coastal populations (as in the case of the red tide in Chile), still seem far removed from a (Northern) land-based perspective, unless the food we eat is affected, as in the case of antibiotics in salmon or organic pollutants in shellfish. It is then that it will become more evident that these regional laboratories or experiments in marine environments could have global repercussions in the future. Moreover, in the case of plastic dispersal, emerging studies exploring freshwater contamination, aerial emissions and soil contamination with microplastics demonstrate that plastic is not merely a problem of marine pollution. Thus, the view of synthetic polymers as a concrete and macroscopic material while they are still in use might dissolve in favour of a more fluid and/or aerial conception of their materiality that takes account of plastics' many afterlives, including as leaching chemicals, or as invisible fine dust emissions that are hugely problematic to us because they permeate the very air we breathe.

Two Forms of Slow Disaster in the Ocean

As outlined in this chapter, plastics do not biodegrade, but rather fragment into smaller particles (micro- and nanoplastics) that will endure for a long period and thus become part of geological deep time: "It is the materialization of the horror of identity, of the stability of form, of a futurity without change" (Davis 2015a, 353). Due to their relationships with chemicals and with biota such as algae, plankton and microbes, microplastics create new habitats and synthetic–biological entanglements, and thus also have an impact on the spread of phenomena like algal blooms. The long-term effects of these transformations within marine ecosystems are only just being investigated, and are at present generally unpredictable. The phenomenon itself, while known within the marine sciences for decades, has only received public scrutiny in recent years.

Due to their ubiquitous existence and mobility in watery environments, ocean plastics might be understood as a form of slow disaster. Although plastics were not introduced with the intention to alter the marine environment and coastal zones, this phenomenon can no longer be dismissed. Moreover, a narrative of a pristine or pure nature still endures: in the present case, a narrative of the ocean before it was touched by plastics. As I have discussed in the chapter, in line with feminist and postcolonial STS and queer approaches, caring for these emerging biological–synthetic entanglements requires other forms that go beyond binaries and strict dichotomies between nature and culture, between purity and toxicity, or between the human and the not-so-human. Representations of environmental disasters still tend to prefer images of "fast violence", such as pictures of mass fish deaths or animals entangled in plastics, which encourage short-term actions that do not address long-term consequences. In this chapter, two cases of slow disasters with still uncertain effects were discussed as speculative ecologies. Due to their speculative temporality and the uncertain temporal and geographical occurrence of their effects, they represent a form of slow violence. Harmful algal blooms and the consequences of industrial aquaculture, plastic pollution and global warming are all emerging matters of care concerning the marine environment and for which responsibility must be taken. For a long time, the ocean was merely regarded as a vast, practically unlimited space, beyond the land and therefore also a large dumping ground. This relationship between humans and sea is hopefully in a state of flux, engaging with multiple modes of knowledge production and newly emerging power structures, including the analysis of which human and non-human actors are included and excluded in the production of the ocean's future.

Notes

1 This research was funded from 2016 to 2017 by the Volkswagen Foundation programme "Original – isn't it? New Options for the Humanities and Cultural Studies", and from 2017 to 2018 by the University of Bremen/German Research Foundation exploratory project "Knowing the Seas as NatureCultures".

2 The singer Marianne Joan Elliott-Said (1957–2003) continued this ironic play on the synthetic by giving herself the stage name Poly Styrene. In an interview, she said: "I choose the name Poly Styrene because it's a lightweight disposable product. [...] Poly Styrene. Plastics. Disposable. That's what pop stars meant to me" (Kennedy 2011). Hence, synthetic polymers and plastics can be considered objects predestined for punk: disposable products that reject a fixed identity and authenticity, and express an emergent way of life.

3 My notion of "speculative" draws on Isabelle Stengers' (2006) discussion of A. N. Whitehead; see also Gabrys 2017, who connects the concept to studies of environmental pollution; the artist Pinar Yoldaş' work on speculative organisms that metabolize plastics is also of interest (Mertens 2014).

4 Names have been pseudonymized with the exception of established scholars I met in the field and whose papers are also cited in this chapter. Quotes from interviews in Chile have been translated into English.

5 Although I do not follow Thomas' separation between the environmental and humanitarian conditions of the crisis here, his ideas are interesting for analysing political mobilization in the field of tension between neglected subaltern regions and Chilean society and government.

6 A recent article on the production of knowledge about the 2016 algae bloom is revealing in several respects (Mascareño et al. 2018). The appendix includes a comparison of controversial arguments from the interviews. One side sees the contamination of the water "as a consequence of pollution produced by the salmon industry", while the other side sees it "as a natural consequence of harmful algal blooms". Similarly, the emergence of the red tide is seen by one side "as a consequence of the dumping of dead salmons", while the other side sees it as the "natural consequence of climatic conditions". Once again, an attempt is being made to divide the world into social and natural factors; but if "global warming" is seen as a natural consequence, then where exactly can a human factor be located?

7 Many people in Chiloé are dependent upon the salmon industry. Gerhard (2017, 733) argues that most people did not blame salmon farming during the crisis of 2007. Aquaculture has also served as an alternative to potato farming, sheep farming and artisanal fishery. With the red tide of 2016, things might have changed. Most people differentiate between mussel farming (*miticultura*), which is less polluting, and the more damaging salmon farming. While the former is mostly the province of local businesspeople, the latter is operated by multinational companies.

8 Moreover, local environmental activists are keen to bring together discourses about the protection of local biodiversity with questions of historical and cultural difference and the symbolic dimension of place. This makes this field of research relevant to political ecology and anthropology. However, these ideas cannot be developed within the limited scope of this chapter.

9 However, there are objections to the public perception that marine animals such as albatrosses only die from ingesting plastics, which is oversimplistic and does not take other factors into account (Liboiron 2021, 106–108).

10 See also the idea of a "multi-species ethnography" (Kirksey and Helmreich 2010).

11 See, for example, some of the essays in Mortimer-Sandilands and Erickson 2010; cf. Gandy 2012 for a critical perspective on queering political ecology.

12 For a critique of the gendered and racialized dimension of these classifications of nature, see Brahinsky, Sasser and Minkoff-Zern (2014).

13 Unsurprisingly, plastic was the pejorative of the anti-gay *Disco Sucks* movement that sprang up at the same time in the late 1970s as a pseudo-rebellious movement that claimed to be fighting for authenticity (Frank 2007).

References

Amaral-Zettler, Linda A., Erik R. Zettler and Tracy J. Mincer. 2020. "Ecology of the Plastisphere." *Nature Reviews Microbiology* 18: 1–13. https://doi.org/10.1038/s41579-019-0308-0.

Arriagada, Nayadeth. 2019. "Adaptive Capacity in Social-Environmental Crisis: The Case of the Red Tide/Salmon Farming Conflict in Chiloé (Chile)." MA thesis, University of British Columbia. https://open.library.ubc.ca/cIRcle/collections/ubctheses/24/items/1.0380286.

Auyero, Javier, and Débora Swistun. 2008. "The Social Production of Toxic Uncertainty." *American Sociological Review* 73 (3): 357–379. https://doi.org/10.1177/000312240807300301.

Bergmann, Sven. 2021. "Dawn of the Plastisphere: An Experiment with Unpredictable Effects." In *Plastic Legacies: Persistence, Pollution, and Politics*, edited by Trisia Farrelly, Sy Taffel and Ian Shaw, 79–99. Edmonton, CAN: Athabasca University Press. https://doi.org/10.15215/aupress/9781771993272.01.

Brahinsky, Rachel, Jade Sasser and Laura-Anne Minkoff-Zern. 2014. "Race, Space, and Nature: An Introduction and Critique." *Antipode* 46 (5): 1135–1152. https://doi.org/10.1111/anti.12109.

Brown, Jayna. 2011. "'Brown Girl in the Ring': Poly Styrene, Annabella Lwin and the Politics of Anger." *Journal of Popular Music Studies* 23 (4): 455–478.

Buschmann, Alejandro, Laura Farías, Fabián Tapia, Daniel Varela and Mónica Vásquez. 2016. "Comisión Marea Roja: Informe Final." Osorno: Universidad de Los Lagos. https://www.directemar.cl/directemar/site/artic/20180619/asocfile/20180619134106/informe_salmones_y_marea_roja.pdf.

Cabello, Felipe C., and Henry P. Godfrey. 2016. "Harmful Algal Blooms (HABs), Marine Ecosystems and Human Health in the Chilean Patagonia." *Revista Chilena de Infectología* 33 (5): 561–562. https://doi.org/10.4067/S0716-10182016000500011.

Cárdenas, Carlos. 2017. "A Un Año de 'La Comuna de Chiloé'." *El Desconcierto.* https://www.eldesconcierto.cl/2017/05/08/a-un-ano-de-la-comuna-de-chiloe/.

Casabianca, Silvia, Samuela Capellaci, Maria Grazia Giacobbe, Carmela Dell'Aversano, Luciana Tartaglione, Fabio Varriale, Ricardo Narizzano, Fulvia Risso, Paolo Moretto, Alessandro Dagnino, Rosella Bertolotto, Enrico Barbone, Nicola Ungaro and Antonella Penna. 2019. "Plastics-Associated Harmful Microalgal Assemblages in Marine Environment." *Environmental Pollution* 244: 617–626. https://doi.org/10.1016/j.envpol.2018.09.110.

Chen, Mel Y. 2012. *Animacies: Biopolitics, Racial Mattering, and Queer Affect.* Durham, NC: Duke University Press.

Davies, Thom. 2019. "Slow Violence and Toxic Geographies: 'Out of Sight' to Whom?" *Environment and Planning C: Politics and Space.* 0 (0), 1–19. https://doi.org/10.1177/2399654419841063.

Davis, Heather. 2015a. "Life & Death in the Anthropocene: A Short History of Plastic", in *Art in the Anthropocene: Encounters among Aesthetics, Politics, Environments and Epistemologies*, edited by Heather Davis and Etienne Turpin, 347–358. London: Open Humanities Press.

Davis, Heather. 2015b. "Toxic Progeny: The Plastisphere and Other Queer Futures." *PhiloSOPHIA* 5 (2): 231–250.

de la Cadena, Marisol. 2015. *Earth Beings: Ecologies of Practice Across Andean Worlds.* Durham, NC: Duke University Press.

De Wolff, Kim. 2014. "Gyre Plastic: Science, Circulation and the Matter of the Great Pacific Garbage Patch." PhD dissertation, San Diego: University of California. http://escholarship.org/uc/item/21w9h64q.

Dris, Rachid, Johnny Gasperi, Mohamed Saad, Cécile Mirande and Bruno Tassin. 2016. "Synthetic Fibers in Atmospheric Fallout: A Source of Microplastics in the Environment?" *Marine Pollution Bulletin* 104 (1–2): 290–293. https://doi.org/10.1016/j.marpolbul.2016.01.006.

Eriksen, Marcus, Laurent C. M. Lebreton, Henry S. Carson, Martin Thiel, Charles J. Moore, Jose C. Borerro, Francois Galgani, Peter G. Ryan and Julia Reisser. 2014. "Plastic Pollution in the World's Oceans: More than 5 Trillion Plastic Pieces Weighing over 250,000 Tons Afloat at Sea." *PloS One* 9 (12): e111913. https://doi.org/10.1371/journal.pone.0111913.

Frank, Gillian. 2007. "Discophobia: Antigay Prejudice and the 1979 Backlash against Disco." *Journal of the History of Sexuality* 16 (2): 276–306.

Gabrys, Jennifer. 2017. "Citizen Sensing, Air Pollution and Fracking: From 'Caring about Your Air' to Speculative Practices of Evidencing Harm." *Sociological Review* 65 (2_suppl): 172–192. https://doi.org/10.1177/0081176917710421.

Gandy, Matthew. 2012. "Queer Ecology: Nature, Sexuality, and Heterotopic Alliances." *Environment and Planning D: Society and Space* 30 (4): 727–747. https://doi.org/10.1068/d10511.

Gerhart, Andrew. 2017. "Petri Dishes of an Archipelago: The Ecological Rubble of the Chilean Salmon Farming Industry." *Journal of Political Ecology* 24 (1): 726–742. https://doi.org/10.2458/v24i1.20963.

Halberstam, Jack. 2005. *In a Queer Time and Place: Transgender Bodies, Subcultural Lives*. Vol. 3. New York: New York University Press.

Hau'ofa, Epeli. 2008. *We Are the Ocean: Selected Works*. Honolulu: University of Hawaii Press.

Iizuka, Michiko, and Jorge Katz. 2011. "Natural Resource Industries, 'Tragedy of the Commons' and the Case of Chilean Salmon Farming." *Institutions and Economies* 3 (2): 259–286.

Kennedy, Channing. 2011. 'Poly Styrene, Multiracial Feminist Punk Icon, Dead at 53'. *Colorlines*, April 26. https://www.colorlines.com/articles/poly-styrene-multiracial-feminist-punk-icon-dead-53.

Kesy, Katharina, Sonja Oberbeckmann, Felix Müller and Matthias Labrenz. 2016. "Polystyrene Influences Bacterial Assemblages in Arenicola Marina-populated Aquatic Environments in Vitro." *Environmental Pollution* 219: 219–227. https://doi.org/10.1016/j.envpol.2016.10.032.

Kirksey, Stefan and Stefan Helmreich. 2010. "The Emergence of Multispecies Ethnography." *Cultural Anthropology* 25 (4): 545–576. https://doi.org/10.1111/j.1548-1360.2010.01069.x.

Knowles, Scott G. 2014. "Learning from Disaster? The History of Technology and the Future of Disaster Research". *Technology and Culture* 55 (4): 773–784. https://doi.org/0040-165X/14/5504-0001/773–84.

Krupar, Shiloh R. 2012. "Transnatural Ethics: Revisiting the Nuclear Cleanup of Rocky Flats, CO, through the Queer Ecology of Nuclia Waste." *Cultural Geographies* 19 (3): 303–327. https://doi.org/10.1177/1474474011433756.

Lambert, Scott and Martin Wagner. 2018. "Microplastics Are Contaminants of Emerging Concern in Freshwater Environments: An Overview." In *Freshwater Microplastics: Emerging Environmental Contaminants?* edited by Martin Wagner and Scott Lambert, 1–23. Cham: Springer.

Liboiron, Max. 2016. "Redefining Pollution and Action: The Matter of Plastics." *Journal of Material Culture* 21 (1): 87–110. https://doi.org/10.1177/1359183515622966.

Liboiron, Max. 2021. *Pollution Is Colonialism*. Durham, NC and London: Duke University Press.

Liboiron, Max, Manuel Tironi and Nerea Calvillo. 2018. "Toxic Politics: Acting in a Permanently Polluted World." *Social Studies of Science* 48 (3): 331–349. https://doi.org/10.1177/0306312718783087.

Long, Marc, Ika Paul-Pont, Hélène Hégaret, Brivaela Moriceau, Christophe Lambert, Arnaud Huvet and Philippe Soudant. 2017. "Interactions between Polystyrene Microplastics and Marine Phytoplankton Lead to Species-Specific Hetero-Aggregation." *Environmental Pollution* 228: 454–463. https://doi.org/10.1016/j.envpol.2017.05.047.

Mardones, Jorge I., Marius N. Müller and Gustaaf M. Hallegraeff. 2017. "Toxic Dinoflagellate Blooms of *Alexandrium catenella* in Chilean Fjords: A Resilient Winner from Climate Change." *ICES Journal of Marine Science* 74 (4): 988–995. https://doi.org/10.1093/icesjms/fsw164.

Martin, Aryn, Natasha Myers and Ana Viseu. 2015. "The Politics of Care in Technoscience." *Social Studies of Science* 45 (5): 625–641. https://doi.org/10.1177/0306312715602073.

Mascareño, Aldo, Rodrigo Cordero, Gabriela Azócar, Marco Billi, Pablo A. Henríquez and Gonzalo A. Ruz. 2018. "Controversies in Social-Ecological Systems." *Ecology and Society* 23 (4): 1–25. https://doi.org/10.5751/ES-10300-230415.

Masco, Joseph. 2015. "The Age of Fallout." *History of the Present* 5 (2): 137–168. https://doi.org/10.5406/historypresent.5.2.0137.

Mertens, Heike Catherina, ed. 2014. *Pinar Yoldas: An Ecosystem of Excess*. Berlin: Argobooks.

Mortimer-Sandilands, Catriona, and Bruce Erickson. 2010. *Queer Ecologies: Sex, Nature, Politics, Desire*. Bloomington and Indianapolis: Indiana University Press.

Murphy, Michelle. 2006. *Sick Building Syndrome and the Problem of Uncertainty: Environmental Politics, Technoscience, and Women Workers*. Durham, NC: Duke University Press.

Murphy, Michelle. 2008. "Chemical Regimes of Living." *Environmental History* 13 (4): 695–703.

Nash, Linda. 2008. "Purity and Danger: Historical Reflections on the Regulation of Environmental Pollutants." *Environmental History* 13 (4): 651–658.

Nixon, Rob. 2011. *Slow Violence and the Environmentalism of the Poor*. Cambridge, MA: Harvard University Press.

Puig de la Bellacasa, María. 2011. "Matters of Care in Technoscience: Assembling Neglected Things." *Social Studies of Science* 41 (1): 85–106. https://doi.org/10.1177/0306312710380301.

Quiñones, Renato A., Marcelo Fuentes, Rodrigo M. Montes, Doris Soto and Jorge León-Muñoz. 2019. "Environmental Issues in Chilean Salmon Farming: A Review." *Reviews in Aquaculture* 11 (2): 375–402. https://doi.org/10.1111/raq.12337.

Sandlos, John, and Arn Keeling. 2016. "Toxic Legacies, Slow Violence, and Environmental Injustice at Giant Mine, Northwest Territories." *Northern Review* 42: 7–21. https://doi.org/10.22584/nr42.2016.00.

Schrader, Astrid. 2010. "Responding to Pfiesteria Piscicida (the Fish Killer): Phantomatic Ontologies, Indeterminacy, and Responsibility in Toxic Microbiology." *Social Studies of Science* 40 (2): 275–306. https://doi.org/10.1177/0306312709344902.

Shapiro, Nicholas. 2015. "Attuning to the Chemosphere: Domestic Formaldehyde, Bodily Reasoning, and the Chemical Sublime." *Cultural Anthropology* 30 (3): 368–393. https://doi.org/10.14506/ca30.3.02.

Star, Susan Leigh. 1990. "Power, Technology and the Phenomenology of Conventions: On Being Allergic to Onions." *Sociological Review* 38 (1_suppl): 26–56. https://doi.org/10.1111/j.1467-954X.1990.tb03347.x.

Stengers, Isabelle. 2006. "Whitehead and Science: From Philosophy of Nature to Speculative Cosmology." Montreal: McGill University. https://secureweb.mcgill.ca/hpsc/files/hpsc/Whitmontreal.pdf.

Subramaniam, Banu. 2014. *Ghost Stories for Darwin: The Science of Variation and the Politics of Diversity*. Champaign: University of Illinois Press.

Thomas, Eric H. 2018. "Crisis and Catastrophe on Chiloé: Collective Memory and the (Re)Framing of an Environmental Disaster." *Cultural Dynamics* 30 (3): 199–213. https://doi.org/10.1177/0921374018795236.

Tsing, Anna Lowenhaupt. 2015. *The Mushroom at the End of the World: On the Possibility of Life in Capitalist Ruins*. Princeton, NJ: Princeton University Press.

Ureta, Sebastian. 2016. "Caring for Waste: Handling Tailings in a Chilean Copper Mine." *Environment and Planning A* 48 (8): 1532–1548. https://doi.org/10.1177/0308518X16645103.

Wong, Alice. 2019. "The Rise and Fall of the Plastic Straw." *Catalyst: Feminism, Theory, Technoscience* 5 (1): 1–12. https://doi.org/10.28968/cftt.v5i1.30435.

Zettler, Erik R., Tracy J. Mincer and Linda A. Amaral-Zettler. 2013. "Life in the 'Plastisphere': Microbial Communities on Plastic Marine Debris." *Environmental Science & Technology* 47 (13): 7137–7146. https://doi.org/10.1021/es401288x.

13 Futures of Plant-Human Mutualism

Science, Technology and Speculative Fiction

Antónia Szabari

Plants are intelligent, and their intelligence is a powerful ontological, technological, ecological force that we human beings need to respect as the model of our ethical action, reckon with as co-producers of Earth's biosphere and atmosphere, and see ourselves as beholden to as cosmic agents. So goes the argument of scholars from different fields whose work has in the past decade revealed the plant as the new universal (Coccia [2017] 2019), a solution to biotechnological problems (Mancuso and Viola 2015), and psychedelic "superpower" (Doyle 2011). In addition, feminist and new materialist scholars in conversation with the science studies of Donna Haraway define plant intelligence in terms of affective and embodied symbiosis (Hustak and Meyers 2012; Tsing 2015). These thinkers share in common the gesture of attributing to plants a cosmic significance as photosynthesizing producers of energy and atmosphere, foodsources that also set in movement an evolutionary process of natural selection, and powerful chemical factories and models for technologies animals do not possess, but they also offer different models for an ethics of relating to plants, for human-plant relationships, and open up different dialogues with the scientific studies that since the 1930s – at least – have attributed intelligence to plants. Plant intelligence is a dynamic notion that at the same time conjures up many implications because of the value our culture (and economies) put on intelligence, intellect, and brainpower. This is why it is not surprising that among scientists it remains a controversial concept,[1] even if a consensus that is slowly but steadily building around certain properties of plants that comprise intelligence: ability to interact with the environment, succeed in attaining a goal, adapt to environment, problem solve, learn from experience, recall memory and learn, make choices (from Lat. "intelligere"), and self-organize, implying a swarm-intellect or organization that emerges from the bottom up (as in some models of insect hive mind or the human brain) rather than from a plan (Trewavas 2016). Feminist scholars and scientists have often insisted on the embodied intelligence of plants. The notion of plant intelligence is rooted in Charles Darwin's hypothesis about a "root-brain" and Nobel laureate geneticist Barbara McClintock's notion of the "thoughtful cell," a broader concept applying to all cells, which she developed from her studies of maize,

DOI: 10.4324/9781003023319-18

tied to McClintock's own foregrounding of her affective relation to plants and cells in the form of "affective ecologies" (Hustak and Meyers 2012, 77). More specific studies have turned to the competitive capability of plants, which scientists characterize as the ability to make *trade-offs*, individuality (the constant presence of variations that makes individual plants behave differently from other plants), error correction, cooperation (exemplified by the involvement of symbiotic organisms with root systems and by those that cheat on the cooperative endeavour), and the use of Volatile Organic Compounds (VOCs) as language (Trewavas 2016). While all these properties of plants can be summed up through the term "intelligence," the term does not define how we are to relate to plants (our cultures have a built-in metaphysical value of intelligence) or how we negotiate the difference between this plant intelligence and the human-specific intelligence (defined as reason or tied to an organ, the brain, considered to be highly evolved) that a long metaphysical and scientific tradition has privileged. Even plant intelligence is haunted by the metaphysical notion of reason, male and Western, insofar as it is enlisted in (neo-)Darwinist accounts of evolution as agent of adaptation and competition. Learning about all or any of these abilities of plants does not necessarily make us more respectful, does not guarantee a change in plant-human relations, and, in fact, there is the possibility that we are interested in plant intelligence precisely because it serves well the scientific biotechnological orientation of our Western economies.

In *Brilliant Green* by Stefano Mancuso and Alessandra Viola, two biologists who advocate for a notion of plant intelligence, assert that plants open up sci-fi-worthy technological possibilities while influencing human ethics and perhaps law and politics as well. They call plants a "universal" form of life, extraterrestrial par excellence:

> They – or rather their most typical cellular organelles, the chloroplasts – are the link connecting the activities of the whole organic world (that is, of everything we call life) with our solar system's energy center. [...] [P]lants have a universal function for life on our planet. Animals don't.
> (Mancuso and Viola 2015, 156)

Mancuso and Viola go on to explain that specific attributes of plants allow them to "open the way" for new technologies such as "plant-inspired robots," "plant-based networks" or "Greenternet," [...] "phytocomputers," new pharmaceuticals, clean energy sources, new man-made materials, and "incalculable unexplored possibilities in the chemical and biological world" (157–158). At the same time, they call for an even more intensive regulation of our dealings with plant life and its even greater inclusion in our legal and ethical order. "[O]n this [...] moral philosophers, molecular biologists, naturalists, and ecologists [...] unanimously agreed [...]: plants cannot be treated arbitrarily. Their indiscriminate destruction is morally unjustifiable" (159). Mancuso and Viola's reevaluation of plant life takes us right into the contemporary situation of plants as *both* drivers of innovation in science and technology, with *universal* importance through their ability to convert solar energy into starches, *and* newly emerging ethical and biopolitical subjects. From their perspective, plants are already robots – that is, their swarm intelligence fits seamlessly with other developments in AI currently underway. Yet the vegetal model of thought which they champion also rehearses many of the classical elements of the liberal subject, especially in its universality and abstraction. Mancuso and Viola's departure from the traditional botanical perspective by defining plants through a networked, non-organ-based notion of intelligence as technologies in the making results in their doubling down on asserting the ethical and legal rights of plants. Yet the politics is where this interpretation of plant intelligence falters. First, by highlighting the technological potentials of plants, all of which are very human-centred, the authors promote nature-based solutions to climate crisis and ecological troubles (both of which they define as problems for humans, their economies, as in the need to predict extreme weather events), thus reformulating plant science in technocratic and neoliberal terms. Here Delf Rothe's (2018) critique is on target. He calls this the "hortomentality" of the contemporary eco-modernist approach, which places a lot of emphasis on technology, considering nature itself as a technology, and remains blind to the violence of the very technocratic and neoliberal processes it advocates. This way of thinking takes for granted the existence of fair government and a more equitable society without addressing the problem of how we get there or acknowledging the social and political dimensions of the climate crisis. It also hastily assumes that the technologies of other species coincide with the telos of human societies and are waiting to be unlocked by us. This account makes us most clearly aware that plant intelligence caters, to some extent, to the expectations of a biotechnological revolution underway, and it is this anticipated importance of biotechnology in the future that necessitates a corrective ethical injunction. The OECD approach to overcoming the fossil energy-driven form of contemporary capitalism is based on the biotechnological use of the regenerative and productive capacities of all biological entities. This approach entails maximizing the productive capacity of plants beyond the methods of the plantation and industrial agriculture through, notably, gene modification. Yet, at the same time, we may be wondering if the idea of a plant-based bioeconomy is also being delineated by these authors, and by others cited above, who are explicitly concerned with our exploitation of plants and the eco- and biosphere they support.

In this essay, I discuss in particular the question of what it would be like to imagine futures in which a plant-based bioeconomy existed and what would be the role of humans in it, by putting in conversation Sue Burke's recent sci-fi duology, *Semiosis* (2018) and *Interference* (2019) and voices in the recent field of critical plant studies. Burke's sci-fi duology offers representations of scientific notions of plant intelligence and, to a lesser extent, of the intelligence of other living organisms and weaves them into an ambitious utopian political narrative, that of a human settler colony on an alien planet that refuses the logic of colonizing, which succeeds in being suspenseful and thought provoking. Burke does not directly engage with plant studies in the social sciences and the humanities and relies instead on setting up an imaginary scenario in which various scientifically-confirmed abilities (i.e., competition, individuality, error correction, cooperation and language) are granted to plants.

In her sci-fi tales, Burke reverses the account of evolution by introducing a plant whose intelligence is superior to that of human beings. At the same time, the human community is also presented as a work in progress (a possibility offered by the multi-generational settler plot line) and the plant emerges to take over the organization of a multi-species community, which is reaffirmed at different points in the plot as a symbiotic one. I compare Burke's narrative of cooperative plant-human (and other species) societies to Donna Haraway's preferred modes of fabulation, in which humans become undone through a sympoiesis, through deliberately plotted "oddkin" relations. These authors (Haraway and Burke) experiment with kinds of "worlding" (to use a recently popular term in the sci-fi and fantasy community) that takes place in messy, troubled places. What Burke achieves in particular, I show, is the problematization of teleological aspirations of the human collective, for which the plant doubles as both an always differing other, troubling human society and world building with a symbiotic approach to making collectives and one capable of imitating or doubling the human telos, ultimately preventing utopian teleological narratives. However, the plants' doubling of human aspirations, techne, is also a disturbing, dark aspect of Burke's fictional superplant, whose alterity threatens to be a mirror of human civilizatory drive, in a way that Burke's novels remain (intended or not) within the plant horror genre. In the two sections that follow, I examine first, the ways in which Burke's superplant shadows metaphysical narratives of reason and technology, then, the novels' indebtedness to posthumanist notions of plant technologies and psychedelics, and, finally, I turn to exploring connections between Burke's sci-fi duology and new materialist fabulation.

The Superplant and the Reversal of Colonial and Technological Modernity

Semiosis makes use of the narrative element of colonizing a new planet (called "Pax" to reflect the colonizers' original utopian aspirations of living in peace with the native species) and imagines the (not-so distant possibility) that life on earth has reached a political and ecological dead end, not the end of the world, but the end of a liveable world, taken collectively for humanity and nonhuman species alike, a situation that calls for change, in order to examine a possible future. By reversing the world-building hierarchies between humans and plants, giving the plant greater control, the novel models the relation between the humans and the plant as an asymmetric mutualism in which the plant is both a symbiotic partner and an enlightened authority, one that knows the world and is eager to know even more, and thus to govern, reminding us that gardening utopias carry the dark shadow of settler colonialism, exploitation and subjugation of human and non-human species (even when the power that claims superior knowledge and mastery is a plant). As Toni Morrison argues in *Playing in the Dark*, the dual experience of poverty, persecution and death in the Old West and the "clean slate" of North America helped create the white American self as settler colonialist, experiencing unprecedented control over enslaved black people (Morrison 1993, 34–37). *Semiosis* reimagines the clean-slate situation of settler colonialism by adding difficulties like increased gravity and murderous plants and turning the colonizing humans into potential subjects of colonization, while also suggesting a resolution otherwise. In an interview Burke briefly reflects on this reversal by saying:

> As I wrote this, I was aware of the history of the colonization of the Americas. We're still living amid the wreckage, and I'm not sure how much we've learned from that disaster. Exploitation remains a common way to interact with each other and with our world.
>
> (Liptak 2019)

Semiosis also rethinks colonial history as one akin to competition in the Darwinian sense and reverses this situation, albeit competition lingers as a dark possibility. In both *Semiosis* and its sequel *Interference*, the practices of cultivation and gardening are revealed as potentially exploitative in spite of the reversal, for giving the plant the upper hand in gardening is an ambiguous gesture that both puts an end to the long human history of exploiting other people, animals, ecosystems, and the planet *and* the possible repetition of that pattern by vegetal means. On Pax, the Pacifists find themselves outwitted by a superbly intelligent and enormous bamboo plant, one that is both rhizomatic and possesses an individual ability of self-expression. The settlers, instead of simply sustainably cultivating Pax nature to feed their colony, as they had expected, find themselves fed and cultivated by the bamboo plant fully capable of communication, synthetizing chemicals that seduce, sustain, sedate, or kill, and thus exerting control over not just the newly arrived colonists but also a larger eco-system, other plants, animals, and fungi around it. The models of symbiosis and mutualism that the Pacifists desire will only become realizable with the help and governance of this superplant, whose symbiotic abilities outdo theirs, a narrative structure that both carries the shadow of dominance and creates new potentials.

Semiosis tells the first 107 years of the colony through the perspectives of successive generations. It starts with an attempt to escape a dystopian near future (ca. 2065) of an ecologically depleted, socially unequal and war-ravaged earth by a group of fifty humans whose aim is to trade it for a new habitat on an alien planet of their (and, eventually, their spaceship's algorithms') choice, where they hope to find a better collective life, in harmony both with the ecosystem of the planet and each other. Both these aspirations turn out to take more than a firm resolution. The Pacifists who settle on Pax are among the lucky few who have the power to fundraise and leave behind earth and find an earth-like habitable exoplanet, which reveals their dependence on earthly technology and material resources, skills, and ideas. The leader of the group, Paula declares:

> 'What human beings and other sentient species bring to the universe,' she said, 'is the ability to make choices, to step beyond the struggle to survive and be the eyes and ears and minds and hearts of the universe. Survival is just the first step.'
>
> (Burke 2018, 29)

The idea of living in peace, moving beyond a neo-Darwinian struggle that capitalistic and colonialist societies have espoused as their guiding principle, is a utopian aspiration of the group that runs into difficulties as interrelational violence breaks out and the group's survival is threatened by loss of technology, lack of resources and predatory species (including predatory plants) on Pax. Paradoxically, they are rescued from becoming colonized by the snow wine, an aggressive native plant, by choosing to cooperate with another Pax plant, the more intelligent snow wine that they name the rainbow bamboo from its ability to communicate with humans through changing colours and, ultimately, "Stevland" to honour one of the members of the group, a musician who died during the long interstellar journey. Stevland is a character in the novel, generally referred to as a "he" (in spite of the fact that it is an androgynous plant), and the only one to continuously appear in this narrative about successive generations. Stevland reminds us of a long tradition of sci-fi narratives that are articulated as patriarchal or colonizing narratives, privileging forms of male, disembodied, exclusionist forms of rationality inherited from dominant discourses of Enlightenment and scientific rationalism. With Stevland taking such a central place, female and racialized bodies are either placed to secondary place or become erased (as is race, through hybridization, in the novels). Race and gender, queerness, as sites of difference that challenge the universalism of Enlightenment thought, disappear from the novel as the successive generations lead to new hybridization between species, genders, and races and ethnicities. The doing away with these historical categories of subjection also means that *Semiosis* and *Interference*, in spite of the decolonial thrust of their narratives, contribute little to the postcolonial, queer, or feminist iterations of sci-fi that are currently being pursued by other authors. What these novels do provide, however, is a good starting point for thinking critically about our view of plants as technology and the concept of plant intelligence as shadowing metaphysical notions of reason and Enlightenment as well as offering post-humanist, and new materialist openings for imagining futures of interspecies intimacies.

Generally, when we think of intelligence, we are primed to think of human intelligence, for a long history of metaphysics tells us that reason is the distinctive feature of humanity (and within it of men, and especially white men). We no longer think of human intelligence as it developed through evolution, as the brain did, from the bottom up, in situations when our ancestors had to make decisions affecting life and death but in terms of morality and science and technology. *Semiosis* places human settlers into life-death situations, to level the gap between them and other species or "nature," while elevating Stevland, the rainbow bamboo (named after its colour scheme but also evoking sentimentalist narratives of human survival), to the rank of superior intelligence. In this scenario, the novel explores a symbiotic cooperative relationship that is (mostly) non-competitive. While the novel places humans who have left a civilization steeped in war technology and exploitive media in an eco-system where they are the prey of other species, it avoids both biological and techno-modernist essentialism, and, instead, reads as a fictional reflection on evolution and technology as each-other's correlates.

French philosopher Bernard Stiegler argues that evolution was suspended with the birth of human techne, including language and simple tools, thus challenging Darwinism that threatens to become the ideology of the techno-scientific era (Stiegler 2008, 176–177). Stiegler's account forms one in a series of philosophical articulations of the indeterminacy between the body and technology, the understanding that technology is not something external or added to the body but is integral to it (following notably Martin Heidegger, Jacques Derrida and Donna Haraway, whose accounts I will not reiterate here). The false Darwinism of techno-science appears in Burke's novels particularly through images of technological wars and war technologies that earthly humans pursue, which become especially prominent in the sequel to *Semiosis*, *Interference*, set in part on Earth. To oppose this neo-Darwinist logic of techno-science, Burke offers a new narrative of evolution, one in which the plant has evolved superior intelligence, is also technological, and is capable of trading technologies with other species (or seizing them). Burke's *Semiosis* is steeped in insights gleaned from recent studies on plant biology, biochemistry, and recent studies and emerging scientific research on plant communication, sensing and intelligence. Making the plant more intelligent is a tongue-in-cheek gesture, for Stevland's intelligence mimics what we know about plants. Plant intelligence throughout the novels debunks techno-Darwinism defined as adaptation and the survival of the fittest because Stevland uses its superior intelligence to suspend competition that characterizes the neo-Darwinist logic of techno-capitalism. His namesake the musician Stevland Barr ties Stevland to human techne, including not only music but also writing, languages, radios, space travel, and drugs (all of which Stevland develops in cooperation with other species). Stiegler defines technics as the pursuit of life by means other than life. For Stiegler, who is a humanist, technology is what constitutes the human, something the body as human coevolves with, and yet always remains foreign to it. While Stiegler does not envision a plant-based techne, one which gives plants agency, he constitutes their being as intelligent. In Burke's novels, Stevland acquires more and more technology, by means of *other lives*, and in doing so mimics the indeterminate relation of the human with technology. Insofar as Burke's fictional super plant learns more from and coevolves with human technologies, it can offer solutions to human problems by eliminating competition but also threatens to supplant the human by becoming, simply, its double.

Plants Get Ideas: The Cosmic Significance of Plants and Psychedelic Technologies

Humanity's exploitative approach to the natural world has been linked to a metaphysical tradition that places human reason on top of a hierarchy and plants on the bottom. Defining technology as not the invention of humanity, serving the improvement of the human, progress or adaptation, but as indeterminately intertwined with the human challenges the progress-driven narrative of modernity, while still referencing a link between the human and techne. However, recent developments in plant biology have contributed to ideas of a plant techne, fully alien to the human, as in the philosophy of Emanuele Coccia, who also underscores plants' universality in contrast to the "parasitic" life of all other life forms:

> To live is essentially to live the life of another: to live in and through the life that others have been able to construct or invent. There is a sort of parasitism, a universal cannibalism, that belongs to the domain of the living: it feeds off itself, without realizing that it needs other forms and modes of existence. As though life in its most complex and articulated forms is never anything but an immense cosmic tautology: it presupposes itself and produces nothing other than itself. [...] As for plants, they represent the only breach in the self-referentiality of the living.
>
> (Coccia 2019, 7–8)

According to Coccia, it is this opening the plant represents towards the universe, the fact that plants "require nothing but the world, nothing but reality in its most basic components: rocks, water, air, light" (8), turning earth into a "cosmic laboratory," which renders plants embodied, material *and* universal, always desiring to "merge with the world" (14). This claim suggests that plants are the necessary technology that is folded into all the living, and plants are moreover foreign or at least cosmic, as they assure that the living is connected with the cosmos. Coccia's main term is the living, rather than the human, and in outlining a plant-based biotechnology, he announces an intentionally non-humanist or post-humanist programme, whose telos is not the survival of the human race. For plants are "the cosmic tumor of humanism" (13). With its cosmic scope it appears as a good companion to sci-fi. For Coccia, plants lead us into a realm of the universal, one that is neither Aristotelian universals, nor Stoic moral reason, not even Kantian moral imperatives, but something non-rational and akin to dance. The plants' leaves, like "flightless birds," he confirms, are immersed in the atmosphere they have also created and create (33). Plant bio-technology is thus one of immersion in a cosmos, and it is akin to dance, music, and other immersive experiences. To other critics, these technologies are psychedelic.

Richard Doyle, author of *Darwin's Pharmacy* has questioned the basic tenets of liberal subjecthood, its universalism (that often masks structured that privilege white, male, Western, human subjects) and rekindled the concept of human immersion in a greater cosmic "life" by way of the early twentieth-century concept of "noosphere"[2] and post- Second World War popular thought on psychedelics, championed, for example, by Aldous Huxley, the Beatniks, and American gurus of psychedelic experience and digital technology Timothy Leary and Terence McKenzie. Doyle's work represents posthumanist reflections on psychedelic technology that consists of rhetoric, language, conceived as poetic and metaphorical, whose creation in human culture he attributes to "plant superpower" (Doyle 2011, 8). Doyle argues, in a performative, personal narrative that draws on the author's own experiences of ritual consumption of ayahuasca, that our artistic, creative life, music, seduction, poetry and ideas have stemmed from the dense evolutionary interconnectedness of our mind and sensations with the psychedelic drugs produced by plants, as well as from colours produced or displayed by plants and animals in the process that Charles Darwin described as sexual selection.

> The plant intelligence that joyfully runs the planet sees human consciousness as a tool for shuffling genes around. The plant complex's aesthetic sensibilities and desires drove early hominid evolution toward its current form and vice versa – we have coevolved.
>
> (Doyle, 254)[3]

Doyle, moreover, following in the footsteps of McKenna, sees the internet as yet another loop in the noosphere, in addition of ideas, art, science, and rhetoric, psychedelic or (in his term) "ecodelic" technology. Doyle's thought largely avoids reckoning with colonial and postcolonial techno-modernity that made possible European poetry, the use of hallucinogenic plants in the West, plant science, and the internet, but, in *Psychedelic White*, Arun Saldanha (2007), who discusses the conscious efforts of white travellers in Goa to transcend the confines of whiteness through dance, trance, music, shared embodied performance and lifestyle choices that reject Western bourgeois lifestyle, admits to the orientalist character and maintains the potential of these cultures to transform the mainstream. Burke's novels take us to a fictional world that seeks to transcend itself through plant technologies and in doing so, we, as readers, are prompted to ask whether the result is another story of colonizing (even with the inversion) or a post-humanist mutualism made possible by the non-human technologies of plants. Burke, who recalls having listened to the sounds of the Nicaraguan jungle when writing *Semiosis*, also conceives the idea of a counter-cultural social project in thinking up the Pacifists' exodus and settling on Pax. Some of Saldanha's observations in *Psychedelic White* can be seen as transposed in Burke's project onto the plot of a sci-fi narrative in which human settlers allow themselves to be seduced by the cultures of other species on Pax. The collective on Pax lives in psychedelic communism whose human and multi-species alliances are made possible by the rainbow bamboo named Stevland who desires to "merge with the world" (Coccia 2019, 14).[4]

The worlds of *Semiosis* and *Interference* are described as fused with deceptive exploitative necro-media (on Earth) and psychedelic ones (on Pax). The narrators of *Semiosis* represent altered consciousnesses produced by the change of generations, the changes in language and, most likely, the psychedelic fruit consumed, first the cantaloupe-tasting fruit of the snow vine, then the much richer and adaptable fruit of the rainbow bamboo, which are catalysts of the worlding in the novel, whose language functions as an always shifting, provisional indexing of a world. The settlers invent names that transform the inchoate sensations of a strange place into familiar plants and animals, "the snow-white vines that are jointed like bamboo" become "Snow Vine," the "spiny caterpillar-like creatures" become slugs and "little land crabs with trilateral symmetry" become deer crabs (consumed by the subsequent generations in stews), "flightless birds-shaped animals with spiny feathers" become "ground eagles," "furry greenhouse-cat sized herbivores" hoping "like springbok gazelle" become fippocats, "green ribbons knobbed with hydrogen" floating in the air become ribbon plants, floating "cactus-spined balloons" become hydrogen cacti, trees with a "bark of cellulose acetate plastic that peeled off in sheets with razor sharp edges" become just one of the many creatures with razor-sharp claws, thorns, or barks, the fungus that causes "hyaline membrane disease, collapsed lungs" is just one of the fungi discovered, the giant kangaroos "with claws like machetes" become fippolions (Burke 2019, 283). Stevland realizes that the Pacifists are intelligent and capable of communication and seduces them with rainbow colours, then with the idea of mutualism. The ecodelic powers of Stevland allow for the thriving of a multi-species community, but this community hinges on often shifting alliances that never guarantee the full "immersion" of the human in a greater totality conceived by Doyle as a feedback loop of semiosis. Stevland most of all strives for the cosmic view, but these pretensions are stymied by his countless daily worries of establishing trade-off's with other species, seducing or convincing them to cooperate, or shutting them out of the benefits of cooperation. Upon the first encounter, the bamboo starts producing alkaloids to seduce the Pacifists, who commit the first murder in order to move from their original colony, where they live in miserable subsistence serving the snow vine, to Rainbow City, an abandoned city whose roofs are made of glass and whose colours mimic those of the bamboo. This city is also the ruins of a former multi-species colony that included the rainbow bamboo and the insect-like Glassmakers (named so by the Pacifists). The Pacifists are equally seduced by the city's architecture and cultural products, to the point of developing affective attachments to the "Glassmakers." Much of the colony's life in Rainbow City is a messy process of symbiotic living with, in which Pacifists benefit from the sensitive and intelligent species they live with, the desire-inducing pheromones of fippolions, intoxicating truffles, playful fippocats, the communications of bats, and the artefacts and relentless difference of the Glassmakers, all contributing to the community. In this multi-species collective, bamboo exercises an outside influence, and, when the Glassmakers return from their nomadic wandering to retake the city, the delicate balance is maintained thanks to Stevland's drugs (opioids and phenylalanine), which allow for an enforced peaceful coexistence between the human-plant-Glassmaker community. In this process, plant-human mutualism allows not only for this multi-species community to thrive but also for Stevland to grow in power and aspiration by using its human and plant "service animals" to extend its control over a larger eco-system. Stevland learns English and learns to communicate with the Pacifists first through Glassmade written language than through written English. Stevland reckons: "I have grown but they have not, and my needs run tandem to theirs in ways that they could never imagine" (Burke 2018, 256). The plant thus reveals to have universal aspirations that are far beyond human or Glassmaker ones, yet he also relies on these species as well as a network of other plants and fungi to unfold its ever-greater aspirations, thus remaining materially limited and potentially exploitative.

Burke's novels offer readers careful narrative thinking about the role of technology in the life of the collective, one that goes beyond the somewhat disappointing tendency to make references to immersion in interconnected systems or loops prevalent in psychedelic literature. In *Interference*, whose story begins in year 2303, a spaceship arrives with new explorers from earth, which in the meantime underwent a new wave of devastation, and two technologies meet. The earthlings, as they are called on Pax, arrive with their radio-based telepathic imaging and communicating devices, imaginary brain enhancing technology. On earth, this technology allows for the maintaining of a public lie, the scapegoating of a woman, only known by the acronym of her name as "NVA," as the alleged perpetrator for a large-scale epidemic that occurred a 150 years before the beginning of the story and a society of spectacle that generates mass delusion, panic and entertainment. The earthlings' consciousness altering technologies allow for the illusion of perpetuating the separation between mind and body, in a way in which the psychedelic technologies of

Pax do not. The materiality of semiosis on Pax is contrasted with the cartesian logic of telepathic communication on earth. *Interference* ends with Stevland communicating with the three bamboo plants he succeeds to send to Earth via messages that take a 110 years to arrive. We anticipate a new colony of ecodelic human-plant mutualism but not without the looming possibility of something going awry; for Stevland's rhetoric of "command[ing] with compassion," as all semiosis, is inherently potentially deceptive (Burke 2019, 304). Is Stevland the fictional incarnation of a universal plant that philosophers like Coccia and Burke assure us transcend our human interests or is he another colonizer, a surrogate of the white man, in disguise of enlightened rationality?

In Stevland's possession, human technology becomes part of a powerful plant-based biotechnology, embodied but carrying universal aspirations. Plants "should be a little scary" Burke says in an interview with Marilyn Dahl posted on the "Shelf-Awareness" Blog in 2018, gesturing towards the favourite trope of plant horror narratives, but also towards the uncertainty and strangeness of plant technology, which leaves the question whether their goals include us, or converge with our civilizatory goals, undecided. This ambiguity, a characteristic trope of the plant horror genre gains a posthumanist twist here. The Pacifists find out in *Semiosis* that the noosphere is not ruled by human thought. The giant bamboo has ideas, thus recalling the plants of popular culture's plant horror genre, like the body snatchers (features as plant-based aliens in two Hollywood movies from the twentieth century) or the comics series *Swamp Thing*, especially as re-created in the 1980's by Allan Moore, who transformed the character of the humanoid plant monster from being a human character, Alec Holland, into a semi-vegetal swamp-dwelling monster to wholly vegetal animate matter with the ability to form the *idea* of being Alec Holland. Burke's Stevland owes something to the plant character of the popular plant horror genre and especially to this gesture of Moore's to grant ideas to the plant-monster (Moore, Totleben, and Bissette 2012–2014). But are the plant's ideas a shadow of human ones as we have known them through history?

Conclusion: The Future of Plant-Human Mutualism

Burke's novels de facto intervene in discussions of critical plant studies via a detour through sci-fi and plant horror to present a constant dialogue about utopian social projects, according to which mutualism is preferable to hegemonic relations, while also giving much thought and narrative power to imagining how mutualism can come about. The best-case scenario is balance between potential foes where they mutually benefit. Stevland is decidedly a scientifically and politically enlightened plant, but its better nature depends on its ability to be dependent on and limited by other species. Burke's is an ecological vision for doing civilization, politics, and social life. This politics is embedded in materiality and limited by it (the lack of metal on Pax thus provides an important limitation to certain technological developments).

Thus, Stevland's ecodelic "superpower" has limits. He is dependent on chemical and material signalling, alliance with other plants and animals, and especially its human "service animals" with whom he builds a narco-communist political community. It is also the universal form of life that aspires to eat the cosmos and become one with it, but this cosmos is revealed to be a material one that provides it constantly with limits in terms of symbiotic possibilities and failures. In attempting to sow its own seeds, to replant its offshoots on earth, Stevland becomes "mother" to plants sent to Earth, thus its androgynous quality comes to light, while in another sense, it reaffirms its nonqueer nature by committing to reproducing itself and spreading its offspring. By bringing language, the materiality of ecodelic media, storytelling, semiosis, into the story, Burke signals towards countering universalist projects through stories of messy interconnectedness, situated in its every detail, materially conditioned, and always bound by the narrative arc. The semiosis that works is material and technology is embedded in a material context.

By making big statements, Stevland also sets himself up to become deceptive. In the meantime, the critters on Pax "work together, play together, live and die together in a multi-species symbiosis" (Haraway 2016, 129). Fippolions seduce Pacifists of both genders with their femurs. Fippocats are playful as are some of the Pacifists. Interspecies affects abound. The way Burke waives plant superpower as both promising and "a little scary" into her narratives is both reminiscent of what Haraway has called "staying with the trouble," that is "not to represent the True, rather to be witnesses for the possibility of other ways of doing what would perhaps be better" (130–131) *and* allows her to stay at a comfortable remove from thinking about Haraway's second trouble, the specifically human responsibility for the ecological crisis we are facing on the planet:

> We must somehow make the relay, inherit the trouble, and reinvent the conditions for multi-species flourishing, not just in time of ceaseless human wars and genocides, but in time of human-propelled mass extinctions and multispecies genocides that sweep people and critters into the vortex.
>
> (Haraway, 130)

Haraway offers her own "Camille Stories," exercises in fabulation in which a human community transforms, post-disaster and through a succession of five generations, into a multispecies collective in which the human is gradually erased, through choices of symbiotic relations with other species and a set of collectively chosen laws and genetic modifications. Haraway's Camille stories enact the revolt against Western, capitalist, anthropocentric life on the planet by erasing the human, but in doing so the storytelling also sacrifices its own tools (characters and narrative arc), while the story remains programmatic, not to say disturbingly unspecific about the tools displayed like eugenics and genetic modification or what permits and makes possible, materially, these technologies. Burke's novels escape the "paralysis of critique," utopian efforts at doing the work of world-making based on representations of the True as Donna Haraway suggests (Terranova 2016). But they also delegate the ultimate responsibility to a superplant.

Burke's fiction avoids the joyous worlding through loss of Haraway's "Camille Stories" in which a human community makes a calculated and technologically assisted effort to disappear into a sympoetic being with other species, but the novels, as they churn through the passing of generations that make the relay, still remain haunted by loss. The reviewer in *Kirkus* noted of *Semiosis* that the otherwise "outstanding" sci-fi story is "hobbled" by its "rushed story structure," in which

> the story's overwhelming scope is also its downfall: readers scarcely have time to register who the colonists are and what's happened during the intervening years before being rushed forward again. Interesting storylines end abruptly, and action scenes, including a monumental battle, feel rushed.
>
> (Kirkus Reviews 2018, 88)

And indeed. Each character and plot established fades away, with Stevland being the only continuous character, who thus signals the greater time scale of plants in a given biosphere and noosphere. While Haraway narrates the gradual and intentional disappearance of a human community through selecting consciously for decreasing human population and interspecies assemblages, Burke has created a sci-fi story in which the accelerated story line allows her to phase out the realist elements of sci-fi narratives, plots and character, especially human ones. Narrative here helps us come face to face with the distant possibility of human civilization, art and technology not serving human goals at all, being incorporated into other, competing and possibly cooperating civilizatory processes, of which the most durable in her novels are those of plants. But also, in Burke's fiction the techne of human(oid) writing, slowly becoming less human, serves as the mnemonic device that creates the continuity of Pax society and also provides the antidote of collective negotiation to the consumerist, industrial, deceptive and cruel media that hold Earth societies in their grip in *Interference* and in our extrafictional present.

> The risk of listening to a story is that it can obliterate us in ramifying webs that cannot be known in advance of venturing among their myriad threads. In a world of anthropozoogenesis, the figural is more likely than not to grow teeth and bite us in the bum.
>
> (Haraway 2016, 132)

Haraway thus emphasizes the instabilities brought to anthropocentric thinking by narratives that crisscross between human and animal stories. *Semiosis* and *Interference* involve the plant, as a character not without anthropocentric qualities, but with the aspiration to transcend them, designed to underscore the radical difference of the plant, and its troubled relation with its double, humans, in the future worlds of ecological fiction.

Notes

1 "Plant intelligence" is controversial in the natural sciences, where the terminology used to describe it is debated, and because it has sparked new vegetal ontologies and metaphysics. Empirical sciences provide us with evidence of plant sensing, chemical signalling, algorithmic life functions, ability to form networks and symbiotic relationships, and photosynthesis, but whether we define these functions as amounting to intelligence is a subject of debates that cut through very notion of intelligence. A good introduction to the scientifically debated concept is provided by Michael Pollan (2013), Stefano Mancuso and Alessandra Viola (2015), and Daniel Chamovitz's (2013). Novels by Jeff VanderMeer's (2014) and Richard Power's (2018) also address plant intelligence, albeit through different aesthetic and narrative forms, weird fiction and realist novel, respectively.

2 The idea that there is a realm of ideas what is otherwise, the realm of intelligence and communication, called the "noosphere" (from "noos" or "mind"), distinct from the geo- and bio-spheres making up the earth, forming a continuous feed-back loop with them, is itself a form of elevating human ideas to form part of a greater whole, according to the creators of this concept two twentieth-century thinkers, a scientist Vladimir Vernadsky and the French Jesuit theologian Pierre Teilhard De Chardin.

3 To some extent, Michael Pollan's journalistic writings have been a forum for expressing similar suggestions about "plant super power" and the ability of plants to shape our desire and our ability to participate in "evolutionary conversation" with plants through acts of gardening and consumption, induced by colours, textures, drugs, sugars and other chemical substances produced by plants. See especially Pollan (2002).

4 I thank graduate students in my Comparative Literature graduate seminar on "Blasted Landscapes and Speculative Fiction" for a close reading of Burke's *Semiosis* with me. To Cynthia Zhang, I owe the term "psychedelic communism". Daisy Reid and Matt Kessler have contributed interpretive ideas on modelling multi-species collectives and narrative structures.

References

Burke, Sue. 2018. *Semiosis*. New York: Tor Books.

Burke, Sue. 2019. *Interference*. New York: Tor Books.

Chamovitz, Daniel. 2013. *What a Plant Know: Field Guide to the Senses*. New York: Scientific American / Farrar, Straus and Giroux.

Coccia, Emanuele. 2019. *The Life of Plants: A Metaphysics of Mixture*. Cambridge, UK: Polity.

Doyle, Richard. 2011. *Darwin's Pharmacy: Sex, Plants, and the Evolution of Noösphere*. Seattle: University of Washington Press.

Haraway, Donna J. 2016. *Staying with the Trouble: Making Kin in the Chthulucene*. Durham, NC: Duke University Press.

Hustak, Carla and Natasha Meyers. 2012. “Involutionary Momentum: Affective Ecologies and the Sciences of Plant/Insect Encounters.” *Differences: A Journal of Feminist Cultural Studies.* Special Issue “Feminist Theory Out of Science” edited by Sophia Roosth and Astrid Schrader. 23 (3): 74–117.

Kirkus Reviews. 2017. “Semiosis: Burke, Sue.” Review of *Semiosis,* by Sue Burke. *Kirkus Reviews* 85 (24): 88.

Liptak, Andrew. 2019. “Sue Burke on Her Next Sci-Fi Novel *Interference*, Colonization, and First Contact: And Why You Should Be Nice to your Houseplants.” *The Verge* (March 29). https://www.theverge.com/2019/3/28/18283790/sue-burke-interference-semiosis-science-fiction-novel-cover-reveal.

Mancuso, Stefano and Alessandra Viola. 2015. *Brilliant Green: The Surprising History and Science of Plant Intelligence.* Translated by Joan Benham. Introduction by Michael Pollan. Washington, DC: Island Press.

Moore, Allan (writer), John Totleben (illustrator), Stephen Bissette (illustrator). 2012–2014. *Saga of the Swamp Thing, Books* 1–6. New York: DC Vertigo.

Morrison, Toni. 1993. *Playing in the Dark: Whiteness and the Literary Imagination.* New York: Vintage. Reprint Edition.

Pollan, Michael. 2002. *The Botany of Desire: A Plant's-Eye View of the World.* New York: Penguin.

Pollan, Michael. 2013. “The Intelligent Plant: Scientists Debate a New Way of Understanding Flora.” *New Yorker.* December 16. https://www.newyorker.com/magazine/2013/12/23/the-intelligent-plant

Powers, Richard. 2018. *The Overstory.* New York: W. W. Norton & Company.

Rothe, Delf. 2018. “Growing Resilient.” Paper presented at the International Studies Association Annual Symposium, San Francisco, April 4–7.

Saldanha, Arun. 2007. *Psychedelic White*: *Goa Trance and the Viscosity of Race.* Minneapolis: University of Minnesota Press.

Stiegler, Bernard. 2008. *Technics and Time 2: Disorientation.* Meridian Crossing Aesthetics. Translated by Stephen Barker. Stanford, CA: Stanford University Press.

Terranova, Fabrizio, dir. *Story Telling for Earthly Survival.* 2016. DVD.

Trewavas, Tony. 2016. “Plant Intelligence: An Overview.” *BioScience*, 66 (7): 542–551. https://doi.org/10.1093/biosci/biw048

Tsing, Anna Lowenhaupt. 2015. *The Mushroom at the End of the World: On the Possibility of Life in Capitalist Ruins.* Princeton, NJ: Princeton University Press.

VanderMeer, Jeff. 2014. *Southern Reach Trilogy.* New York: Farrar, Straus and Giroux.

Index